ARDUINO FLYING PROJECTS

Robert J Davis II

Arduino Flying Projects
How to Build Multicopters from 100mm to 550mm.
Copyright 2017 by Robert J Davis II

I know that there are lots of other books, articles and web sites on the subjects of Drones, Quadcopters and Arduino's. I have felt compelled to add my book to the growing collection. Hopefully I will provide the answers to all of your questions located in just one place. You can find your answers on the Internet but you will spend many wasted hours looking for them.

This book will start with covering the basics. Then it will show how to build quadcopters, racing quads and even a hexacopter. Last of all it will show some of the many available options.

As always, the safe construction and operation of the projects in this book is the sole responsibility of the builder. These flying projects can easily crash, self destruct and otherwise cause a lot of damage. Be very careful around the spinning propellers! These flying machines can cut off a finger and otherwise do serious bodily harm.

Back around the turn of the century I made some predictions for how things will change this century. I predicted that the telephone, PDA (Personal Digital Assistant) and computer would all be combined into one device. Next, I predicted that power, telephone and cable wires would all go away. Many people will even have their own power generators. Last of all I predicted that we would all fly in flying cars. I was laughed at on the last prediction. But if you take one of these quadcopters found in this book and scale it up to "people" size then you have a flying car.

GPS and autonomous flight controllers will make it a possibility. In fact they could work like a taxi, arriving on demand and flying you to your destination totally controlled by GPS and a computer. The flight control processor makes "Fly by wire" possible as there is no "mechanical" method of steering. GPS makes possible "highways in the sky" thus directing the flow of traffic. The technology that is used for the autonomous cars can also be applied to flying devices. So, not only is it possible to build a flying quadcopter, we could also build a flying car. The lithium battery helps to make it all possible by providing a light weight power source.

Contents

1. Background..5
       Sizes   Brushed vs. Brushless

2. Special Tools. 7
       Soldering Iron   Parts Holder
       Magnifying Glasses   Prop Balancer

2. Lithium Batteries and Battery eliminators.......................... 10
       LiPo Batteries   LiPo Charger
       Make Your Own   BEC/Regulators

3. Motors and Electronic Speed Controllers........................... 17
       How Servos Work   Servo Test Program
       Getting Spare Parts   Calibrating the ESC's

4. Remote Control Transmitter and Receivers........................ 25
       Four Channels   Six Channels
       Arrangement of the Controls

5. The F450 Starter Kit.. 28
       Parts in the Kit   Kit Assembly

6. Multiwii, Uno and Nano Flight Controller.......................... 38
       MultiWii   MultiWii Pro
       Accelerometer/Gyroscope Modules
       Connection Diagram   Uno/Nano Adapter
       Software Setup   Config.h
       Calibrating the ESC's.

7. APM Flight Controller 53
       GPS   Power Monitor
       Mission Planner

8. Hobby King KK Flight Controllers................................... 58
       + or X Configuration   USB Programmer

Flashtool

9. The ZMR 250 Racing Quad... 62
    2300KV motors   5" Propellers
    4in1 ESC   1500mA Battery
    Frame Parts   Assembly

10. The F550 HexaCopter.. 69
    Frame Parts   Assembly
    Flight Controller   UFO Style Cover
    Flight Controllers

11. Brushed 105 mm Quad... 75
    The parts needed   The USB interface
    Multiwii Setup

12. Some Video Options.. 79
    Dash cam   Sports Camera
    Micro Cameras   Mounting Cameras
    Cameras with built in Transmitters
    External Transmitters   On Screen Display
    Power Monitors   VR Goggles
    Pan/Tilt Gimbal

13. Optional upgrades... 86
    Landing Legs   Better Landing Skids
    Spare Propellers   Propeller Guards
    Battery Alarm   Running Lights
    Motor Extensions

14. The Future of Flight.. 94
    Tilt Rotors   Ionic Lifters

15. Definition of Terms... 97

Bibliography.. 101

Chapter 1

Background

You can buy an assembled and tested quadcopter for about the same price as you can build one. However, building a drone is lots of fun and you will learn a lot in the process. But you will likely waste some money as well.

There are officially four different sizes of drones:
1. Less than seven ounces is called "Nano"
2. Seven ounces to 4.5 pounds is called "Micro"
3. 4.5 to 45 pounds is called "Mini"
4. 45 to 330 pounds is called "Small"

In reality you will likely be working with only two types of quadcopters. The smaller, hand held sized units, have what are called "Brushed" type motors and do not use three wire ESC's (Electronic Speed Controls).

The bigger quads use "Brushless" motors and ESC's. The ESC takes a servo signal and converts that to three phase AC to power for the motor. The ESC's actually use an Arduino like processor and hence they can be reprogrammed. The brushed motors have their speed varied by changing the voltage that is applied to them. The brushless motors have their speed varied by the frequency that is applied to them.

There are several problems with the brushed type of motors. They use brushes to convert the DC into AC to make the motors rotate. Brushes have a limited life as they will slowly wear away as they run. The second problem is that it is very difficult to control the rotating speed by the voltage that is applied to them. There is no direct relationship and no guarantee that all four motors will rotate at the same speed with the same voltage applied to them. For those reasons the brushless motors are preferred in a quadcopter.

This next picture shows a 100 mm brushed quadcopter on the left, then a ZMR 250 racing quad, then a F-450, and then a F-550 on the right. The number in their names refers to the greatest distance between their motors in mm.

As you can see the size varies dramatically from the smallest to the biggest. Generally, the smaller ones are for training or racing and the bigger ones are for photography or carrying things.

Here are some rules for learning to fly your multirotor:

1. Start small because you will crash.
2. Practice a lot because you will crash
3. Even with a lot of experience you will crash.
4. Every time you take off you run the risk that you will crash.
5. Carry lots of spare parts because you will crash.
6. If I failed to mention it, you will crash.

Chapter 2

Special Tools

There are some special tools that you will need to assemble your flying machine. First of all, many of the circuit boards that are used in drones will come without their header or connectors. For soldering them you will need both a fine tip and also a medium tip soldering iron. The medium tip will be needed for soldering the larger gauge wires to the ESC's and power distribution board. I actually had to buy a fine tip for my soldering iron for soldering the .1 inch spaced connectors without bridging the connections.

To solder the "bullets" (connectors) onto the motor and ESC wires you will need what is sometimes called a "third hand" or a "parts holder" like what you can see in the next picture.

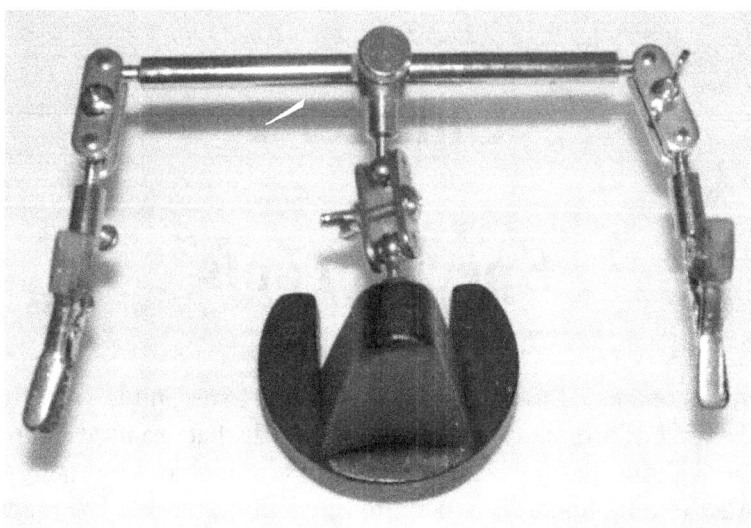

Another thing you might want for the fine soldering is some magnifying glasses. Mine are a little old as the newer ones have a LED light built in.

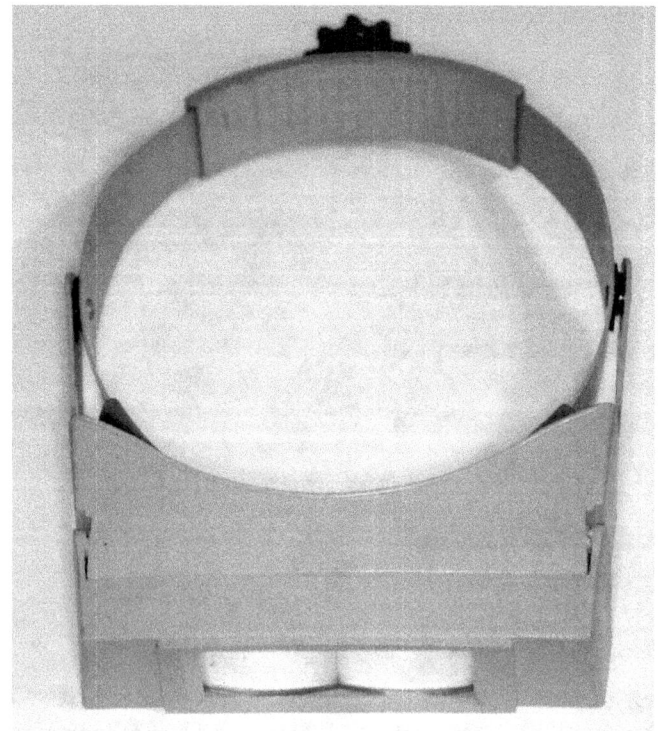

Some people swear by using what are called "prop balancers". In my experience good propellers are usually very close to being balanced when you get them. In case you want to check and adjust the balance, one of these might come in handy. They are not very expensive and do not work on three blade props unless you buy a larger model. You can add some tape to a blade to balance a prop that is out of balance.

You will also need a volt meter that does not have to be an expensive model and common hand tools like needle nose pliers, small wire cutters, jeweler screwdrivers, several small Allen wrenches, several sizes of heat shrink tubing, Velcro, double sided foam tape, and lots of wire ties.

Chapter 3

Lithium Batteries and

Battery Eliminators

Back In the mid 90's I worked for Caldwell Machining in Emporium Pa. Mr. Caldwell had purchased the old Sylvania plant. There was a room in that plant that was to remain untouched forever. It was the lab where I was told that the lithium Ion battery was invented. However, there were many people all around the world that worked on developing the lithium Ion battery.

Most flying machines use lithium polymer batteries. They are called "LiPo" for short. They are safer than the earlier batteries as they do not catch on fire and explode as easily. They might still explode if they are overcharged! That is because it uses a gel rather than a liquid. The discharge voltage is about 3 volts and fully charged they are about 4 volts per cell.

A three cell LiPo battery is the most common model used in a quadcopter. Because all three cells need to be charged separately a "balanced" charger is used. Usually a current rating of 1500 mAh to 2500 mAh is used. That number says that the battery can deliver 1.5 Amps for one hour. But it can also deliver 15 amps for 1/10 an hour or 6 minutes of flight. Some batteries are even rated for up to 30 amps for a short time.

So then you will need a three cell 11.1 volt preferably 2.2 Ah (Amps per hour) 20 to 30C battery pack for a power source for the F450 or F550 Quadcopter and Hexacopter. 2200mAh is the same as 2.2 Ah, and though they are rated for 2.2 amps for 1 hour they can deliver 22 amps for 6 minutes of flight. They run about $15-$20 each. For a longer flight time you can use two identical batteries in parallel.

Here is a typical LiPo battery for sale on eBay for about $19.

Powerextra 11.1V 2200mAh 30C Replacement Battery Pack For RC Airplane RC Helico
(302245660950)

Estimated delivery **Mon, Apr 10**

Along with the battery, but not part of the Quadcopter itself, you will need a three cell balanced battery charger. A two cell or 2S lithium battery is 7.4 volts and a three cell or 3S lithium battery is 11.1 volts. These chargers charge the three cell battery like it was three independent batteries.

B3 AC 100~240V 2S-3S Li-po Balance Charger For 7.4-11.1V RC Battery USA UB
(122270256744)

Estimated delivery **Mon, Apr 10**
Tracking number: 9374869903501870740608

You are better off purchasing a three-cell battery, but you can buy single cells and then wire them together to make a three cell battery. You can do that in one of two ways. One way requires a balanced charger and the batteries are permanently soldered together into one large battery. Messing with Lithium batteries is dangerous as they can explode or catch on fire.

The other option is to charge the batteries separately but then plug three of them into a wiring harness to form a 12-volt battery. The most practical use for such a homemade battery is to run the video camera, lighting and "auxiliary" circuits. You want to have a really good battery for running the motors.

The next picture shows a homemade battery that has three cells and requires a balanced charger. Unfortunately, I had accidently soldered a male connector on the battery and had to change it to a female connector while it was "live". The battery connector should always be female to reduce the risk of it accidently hitting something metal and shorting out. It can melt some metal objects like the rings on your fingers. Do not attempt this unless you really know what you are doing. Shorting a battery can cause an explosion!

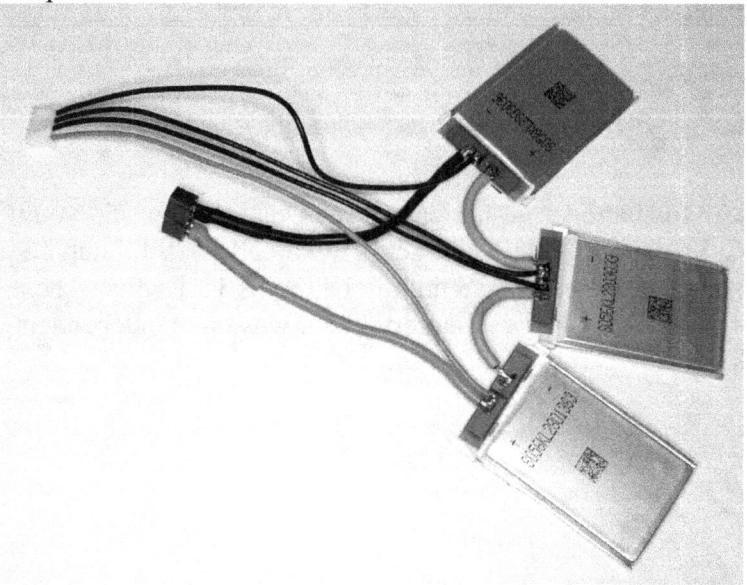

The next step would be to cover the end of the batteries in electrical tape. Next fold them in a "N" pattern and then tape all three batteries together making one three cell battery. The schematic below is for the 3S LiPo battery. The two left wires are red in the schematic.

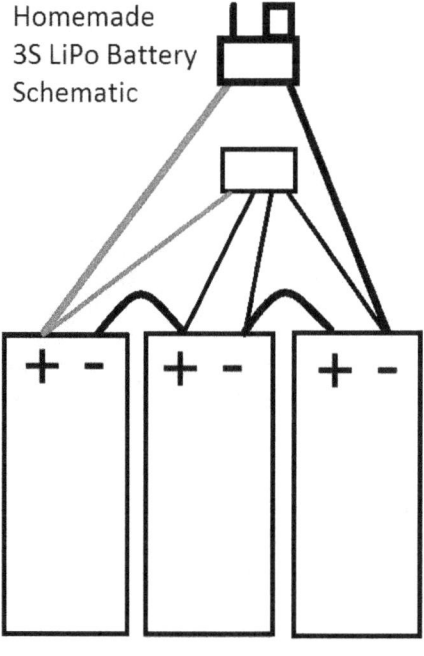

You can also just solder individual connectors onto each battery as seen in the next picture. After soldering the wires, the battery end is wrapped in electrical tape or heat shrink tubing.

Then three of these smaller batteries can be plugged into a serial wiring harness to make a 12-volt battery. Remember that the connectors and the wire size limits the current and this kind of battery should only be used for auxiliary functions like video and LED's.

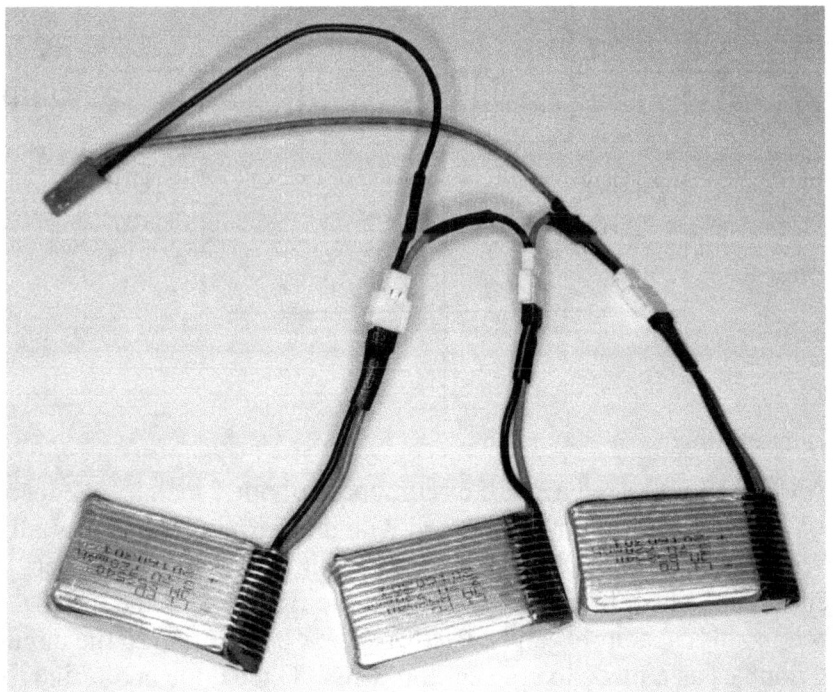

Battery Eliminators (BEC) or voltage regulators can take one voltage from a battery and give you another voltage. You can make a simple 5-volt BEC power supply with a LM7805 or with a AMS1117-5V.

The LM7805 can have the parts directly soldered onto it. Then the assembly can be covered in heat shrink tubing or it can be bolted to a heat sink to handle more power. The metal tab with a hole in it is ground so it can be fastened to any metal piece to help absorb some heat.

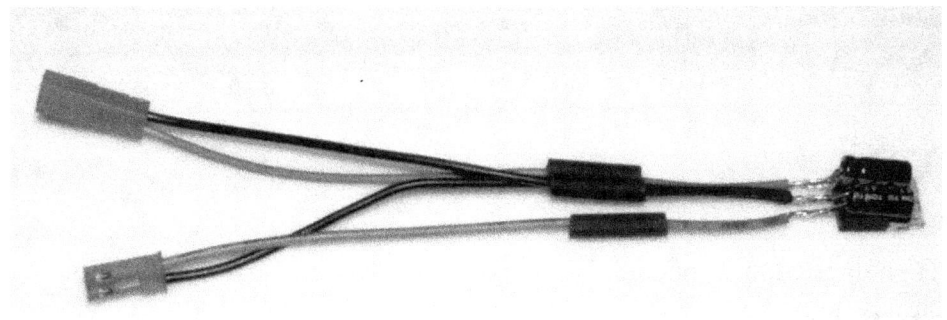

Here is the LM7805 BEC schematic diagram.

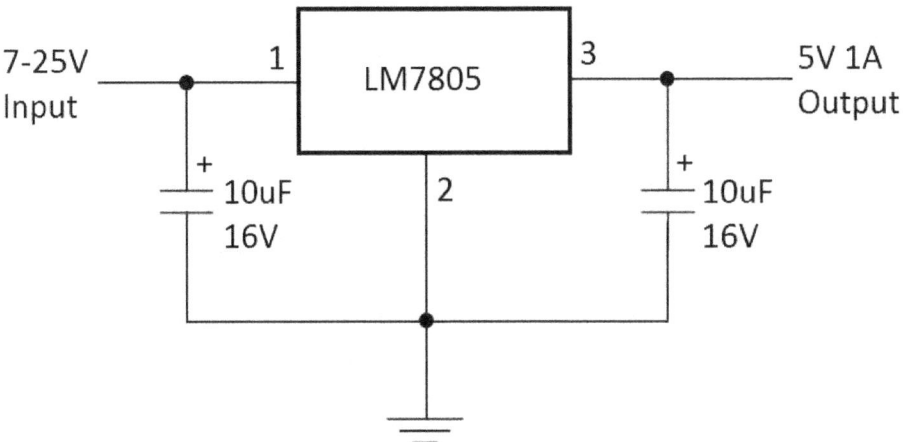

The AMS1117 will need a small circuit board about 1 inch by 1/2 inch in size. You can cut a "T" into the board with an sharp knife and a small screwdriver. The middle lead can be cut off, it is the same as the tab, the power output. The left lead is ground and the right lead is the input. Surface mount capacitors can be used and their connection to the circuit board doubles as a place to connect the wires. Once done and tested the entire board is covered in heat shrink tubing.

The traces on the circuit board can be made with an exacto or razor blade knife and a fine flat tip jewelers screwdriver. Cut out a "T" pattern by going over the area to be removed several times with the knife. Then use the jewelers screwdriver to scrape out the copper pieces. This next picture shows a AMS1117 based voltage regulator or BEC. It can be covered in heat shrinking tubing to make it safer and to look neater.

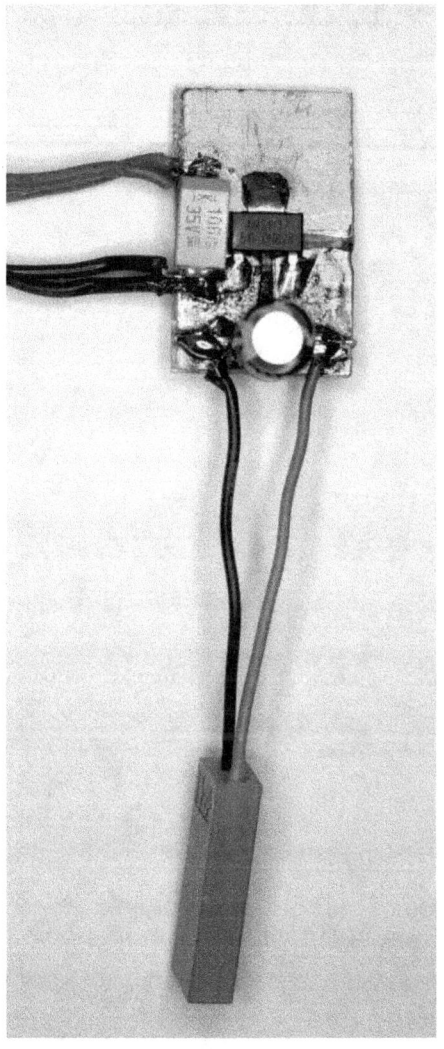

When I first tested the AMS1117 BEC it went up in smoke after only a few seconds of operation. The problem was that surface mounted capacitors have the stripe on the positive end. All other capacitors have the strip on the negative end. I had soldered the capacitors in backwards.

This is the AMS1117 schematic diagram, note that the order of the pins has changed from the LM7805. Pin 2 is the middle pin and the tab. It is easier to solder the tab and bend the middle pin out of the way.

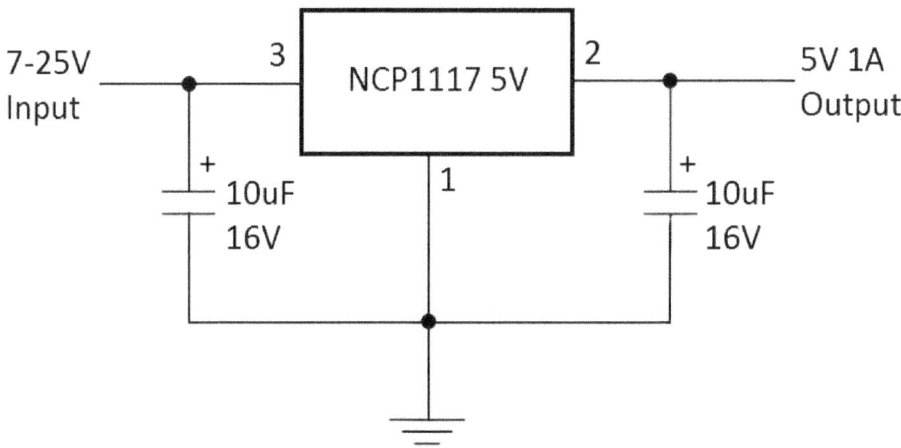

Chapter 3

Motors and

Electronic Speed Controllers

How do servos work? Servo motors must be understood in order to work on quadcopters.

The remote control transmitters, receivers, and ESC's all use "servo" type control signals. That is because the quadcopter evolved out of other flying planes that used servos to control the motors and the flaps. Servos are motors that have a built-in motor controller and a "feedback" loop. Basically, a variable resistor or other device monitors the motor's position. This position information is then "fed back" to the built-in motor controller.

The motor controller takes commands from a processor and then the controller matches up the current motor position with what position the controller was told to move the motor to. Usually the commands are in the form of "pulse width modulation" (PWM) where the width of the pulse tells the controller where to position the motor.

Servo's typically come in several sizes, as can be seen in the next picture. There are normal sizes of servos on the left side of the picture and several micro sized servos on the right side. You can also see that most servos come with a three-pin cable and connector. They carry ground, power, and the PWM signal.

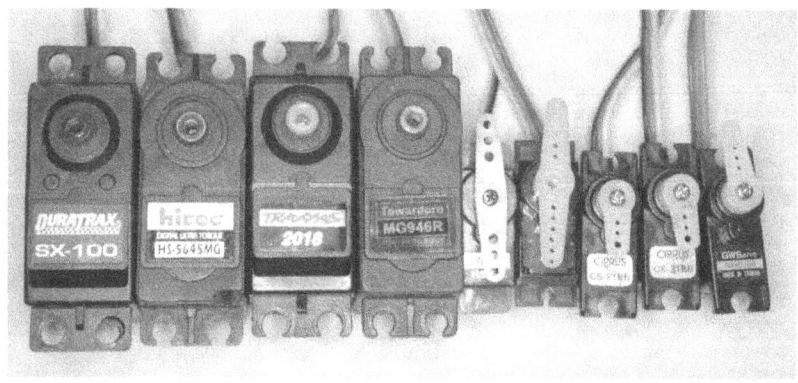

In the case of pulse width modulation commands, usually a pulse width of one millisecond tells the controller to move the servo to zero degrees. A pulse width of 1.5 milliseconds tells the controller to move the servo to 90 degrees. A pulse width of two milliseconds tells the controller to move the servo almost completely open or 180 degrees.

The servo "home" position is supposed to be at 90 degrees. While using a servo test program, with some cheap servos found on eBay, I discovered that a .5 millisecond pulse results in zero degrees of rotation and about 2.0 millisecond pulse resulted in 180 degrees of rotation. The exact servo position relative to pulse width seems to vary between servos. In any case there is also a 20 millisecond delay between each of the control pulses.

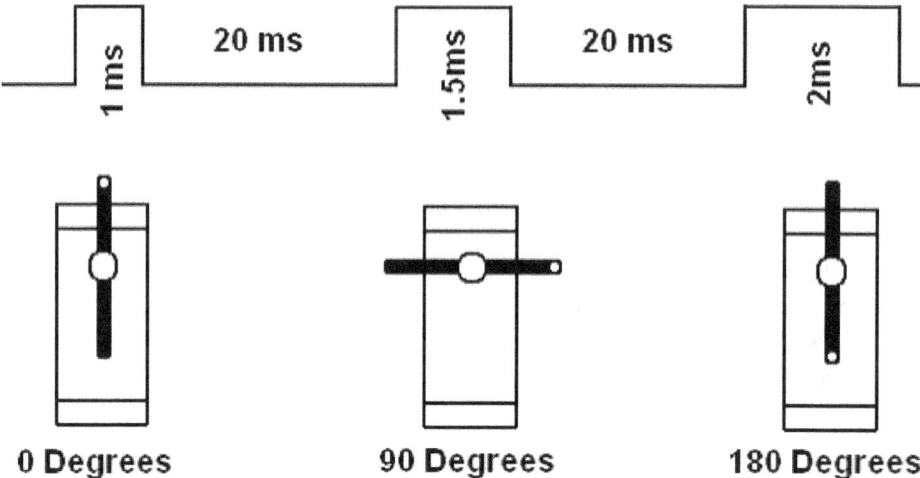

The above chart shows the pulse width and the corresponding position of the servo motor. One millisecond (ms) is "1000", 1.5 ms is "1500" and two ms is "2000" on some flight controller programs.

Some of the advantages of servo motors include that the power source does not have to be switched on or off or otherwise controlled. The power to the servo motor can always be left "on". A PWM output pin of the Arduino can directly control the servo, no driver circuit or transistor is needed, because there is a driver inside of the servo.

Servos make great proportional valve controllers because you can vary a valve from off to full on. For instance, if you want to water your plants automatically and you want the rate of water flow to be adjusted according to the humidity, it can be done with a servo.

This next picture is of the Arduino servo motor test setup that can be used for a servo or ESC test program. You could plug in an ESC with a brushless motor instead. The ESC will require a 12-volt power supply to power the motor. The test program would then vary the speed of the motor according to the position of the variable resistor on the Analog 0 input.

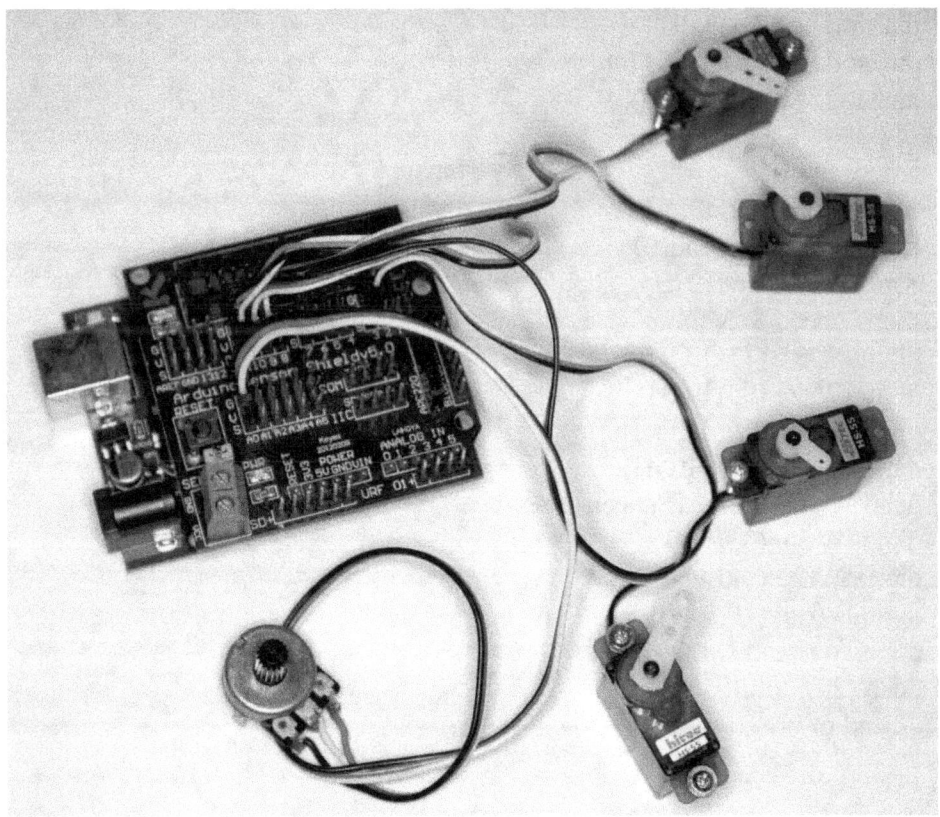

Here is a sketch to demonstrate the operation of up to four typical servo motors or ESC's. ESC's will not turn on until the variable resistor is turned up 1/2 way. That is because the coils in the motor are a "short" at lower frequencies or speeds. The output varies with the value of the input from a variable resistor that is connected to A0. This demo does not use the Arduino servo library.

```
// Servo or ESC demonstration program
// By Bob Davis
// July 10, 2016
// Program to run motors on a quad copter
// According to variable resistor on A0
// ESC's are on pin D3, D9, D10, D11
int Rspeed;
int Sspeed;
void setup() {
// The servos on pins 3, 9, 10, 11
  pinMode (3, OUTPUT);
  pinMode (9, OUTPUT);
  pinMode (10, OUTPUT);
  pinMode (11, OUTPUT);
}
void loop() {
// Analog value is 0 to 1024
  Rspeed=analogRead(0);
  Sspeed=(Rspeed*2);
// Send speed to Motors
  digitalWrite (3, HIGH);
  digitalWrite (9, HIGH);
  digitalWrite (10, HIGH);
  digitalWrite (11, HIGH);
  delayMicroseconds(Sspeed);
  digitalWrite (3, LOW);
  digitalWrite (9, LOW);
  digitalWrite (10, LOW);
  digitalWrite (11, LOW);
  delay(20); //milliseconds
} // End of program
```

Getting Spare Parts

If you want to buy spare motors they are available. Here is where I purchased two more with the ESC's and propellers. The motors cost about $10 each with the ESC. I bought these as part of the upgrade to a hexacopter. To be honest these are not the best motors. The main issue is that the props keep coming loose. The better motors have a threaded shaft for the props that are threaded to match the direction of rotation.

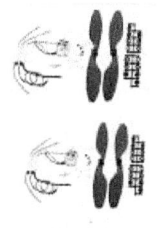

2X A2212 1000KV Brushless Motor w/30A ESC 1045 Propeller for DJI F450 F550 O1B1
(162174110021)

Estimated delivery **Mon, Jun 12**
Tracking number: 74899991206459964606
This item has been shipped.

The additional motors did not come with the bullets to make changing the motors and ESC's easier. They are also available on eBay. They run about ten pairs for $5.

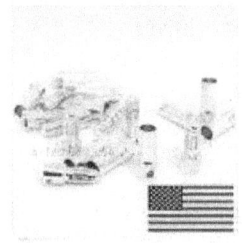

USA 10 Pairs 3.5mm Gold Bullet Connector Plug Male & Female for RC ESC Motor
(112414171689)

Delivered on **Sat, Jun 10**
Tracking number: 9400109699939012282296
This item has been shipped.

For motors I would recommend motors that do not have the problem with the props always coming loose. These motors have the threads built in and the threads are reversed for CW operation. I bought six motors with ESC's for about $70.

6x 2212 920KV Brushless Motor + 6x 30A SimonK RC Brushless ESC With BEC F F450
(112028178721)

Delivered on **Wed, Jun 07**
Tracking number: LK257911265CN

Calibrating the ESC's

This is an important step that I ignored for a long time. My first quad flew fine but I had to readjust the trim on the remote. Then, after a crash, one motors stopped working and the ESC just kept beeping. My second drone only had 2 out of 4 motors that would spin up properly. Then my Hexacopter only had 3 out of 6 motors that would respond properly. The solution was to properly calibrate the ESC's.

1. Connect the ESC to be calibrated to the RC receiver Port 3 (throttle).
2. Turn on Remote Control and turn throttle all the way up.
3. Power on the ESC or the entire drone (Connect the Battery).
4. Wait for all of the beeping to stop.
5. Lower the Remote control throttle down to zero.
6. Wait for all of the beeping to stop.
7. Power down the ESC (Disconnect the battery).

Do this calibration procedure for all of the ESC's on the drone. Once that is done they will all respond the same for the same input pulse width from the controller. Some flight controllers have a built in procedure to calibrate the ESC's.

You can build an adapter to calibrate all the ESC's at the same time. All you need for the adapter is some header strips and then bridge them together like in the next picture. That way the throttle control can be connected to all of the ESC's at the same time.

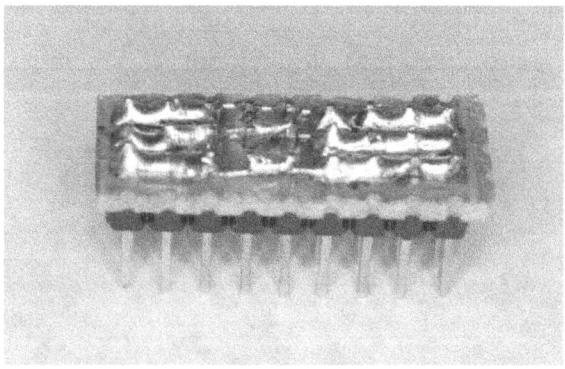

When one ESC would not work I took it apart. On one side of the ESC there is the processor. There are driver transistors to convert the five volt logic to 12 volts. There are copper buss wires to deliver a higher current.

On the other side of the ESC there are the 6 power FET's and two 78M05 voltage regulators. On the left there is a filter capacitor to try to reject any noise that is on the power lines.

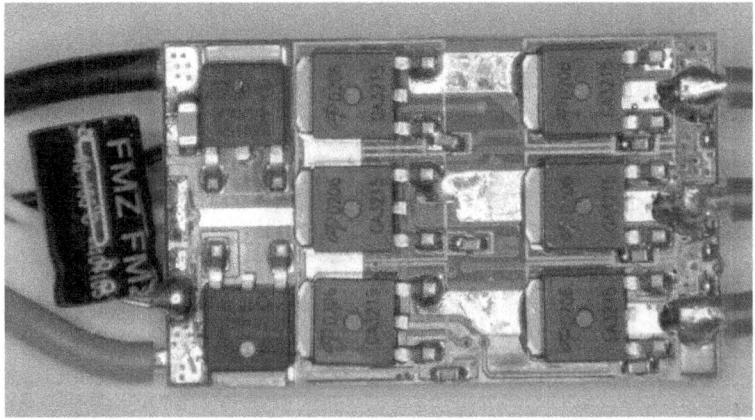

An important note about the multicopter motors is that their mounting holes are not symmetrical. The hole spacing is 16mm one way and 19mm the other way. Also, the length of the screws is critical. If too long of a screw is used it will damage the internal wiring. Ideally a screw should go in three to five complete revolutions. Some people use lock-tight on the motor mounting screws.

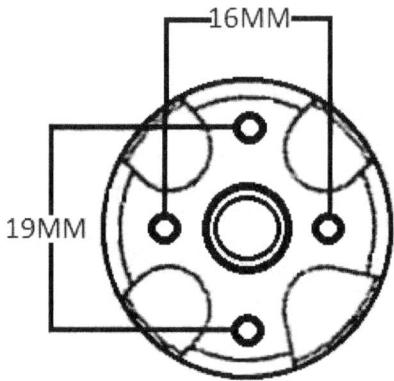

The insides of the motor consist of several coils wound on metal stators that form electromagnets, that part is called the Stator. There are several magnets that are fastened to the rotating part of the motor, that is called the Rotor. The next picture shows the insides of a motor.

The numbers that are on a motor tell you the size and speed of the motor. For instance, having "2212-1000Kv" on a motor indicates that the motor is 22mm round (about 1 inch) by 12 mm tall (1/2 inch, only counting the stator) and does 1000 RPM per volt. A 1000Kv motor will do 12,000 RPM under no load with a three cell LiPo battery.

Chapter 4

Remote Control

Transmitters and Receivers

A few years back Remote Control (RC) transmitters used 72 MHz crystal controlled transmitters and receivers. Now with the advent of digital tuning you can select the frequency on the fly. Below is a picture of a 72 MHz receiver. The control crystal is on the left side at the front.

This next picture is a modern digitally tuned six channel receiver.

For the radio transmitter remote control, you will need a transmitter and a matching receiver. This remote can be the same for a micro, for a full-size Quadcopter or even a hexacopter. A cheap four channel "starter" unit is about $35 from China.

Flysky 2.4G FS-T4B 4 CH Radio Control RC Transmitter Receiver NEW X0S0
(132035970309)

Delivered on **Tue, Mar 14**
Tracking number 61299991206460848401

After a few flights you will likely want to upgrade to a six channel remote. The six-channel unit gives you the ability to remotely change the flying mode of the quadcopter, but it will cost around $50. This model came with female headers because it is designed to plug into an APM and it sits right on top of the APM. The additional channels can control the camera pan and tilt or change modes to auto-level or even signal the drone to "fly home".

Flysky FS-T6 2.4GHz 6CH Transmitter Radio *Note Receiver Pictures & description
(332188362539)

Estimated delivery **Sat, Jun 10**
Tracking number 9405509699939001075550

If your quad tilts during lift off it is likely that your trims are not set correctly and it is telling the quad to turn in a direction. You might have to fine tune the trims again. You can also adjust the trims while connected to a laptop to be able to visually center the trims.

On the remote-control transmitter, the left up/down control is the throttle, it is the only control that is normally at "zero". The others all return to center. The left lever's right/left function is to tell the drone to rotate to the right or left. The right up/down control is to move forward and backwards. The right lever right/left function tells the drone to fly to the right or to the left.

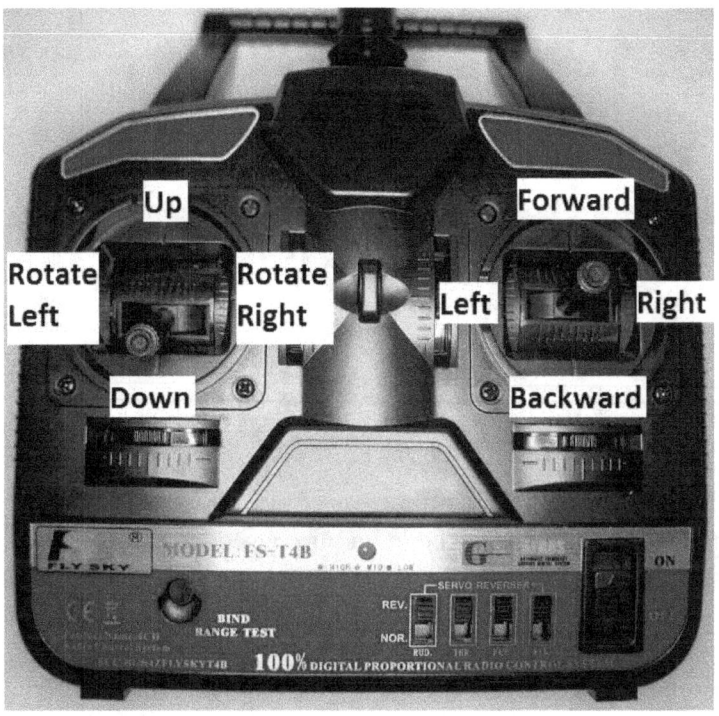

This next picture shows how the controls on the remote control line up with the channels of the receiver.

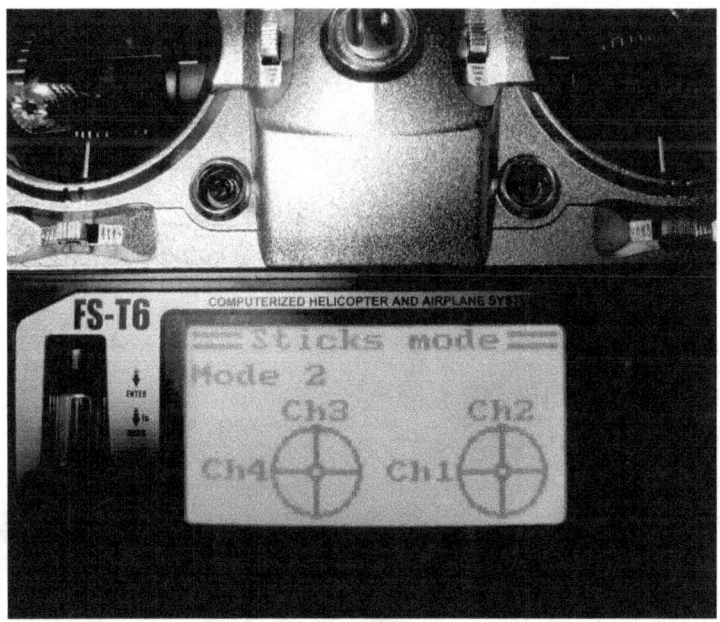

Chapter 5

F-450

Starter Kit

Most of the parts that you need can be purchased on eBay, but you will have to know exactly what to look for. I picked up the very popular DJI F450 DIY (Do It Yourself) frame kit for a little less than the normal selling price. It included the frame, the ESC's, the motors, the propellers and even some tools that are needed for assembly. This kit normally runs about $65 from China. Sometimes these kits are called "ARF" as in "Almost Ready to Fly".

Let's take a closer look at some of the parts that came with the F450 (Flame Wheel 450) partial quadcopter kit.

First, there are the four legs and the top and bottom circuit boards. These parts make up the mechanical frame of the quadcopter itself. They can be seen in the next two pictures.

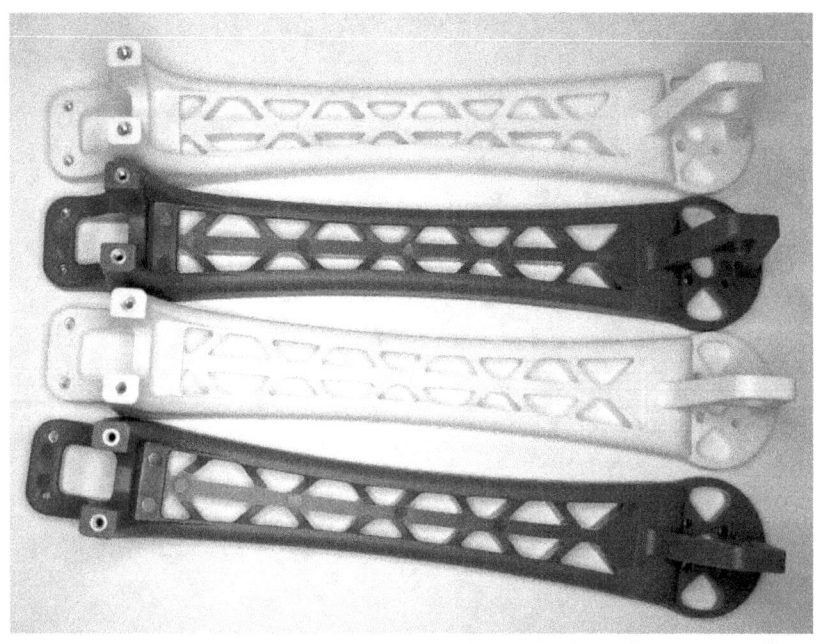

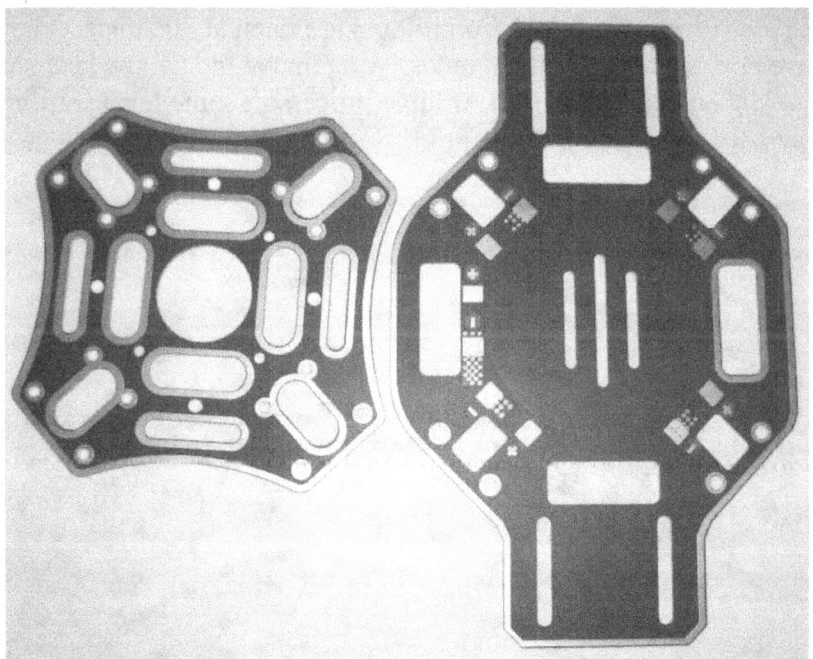

Next there are the electrical parts that came in the kit, the motors and the ESC's. As you can see I already started soldering the bullet connectors onto the motors and the ESC's. The four brushless motors are model number A2212/13T 1000KV

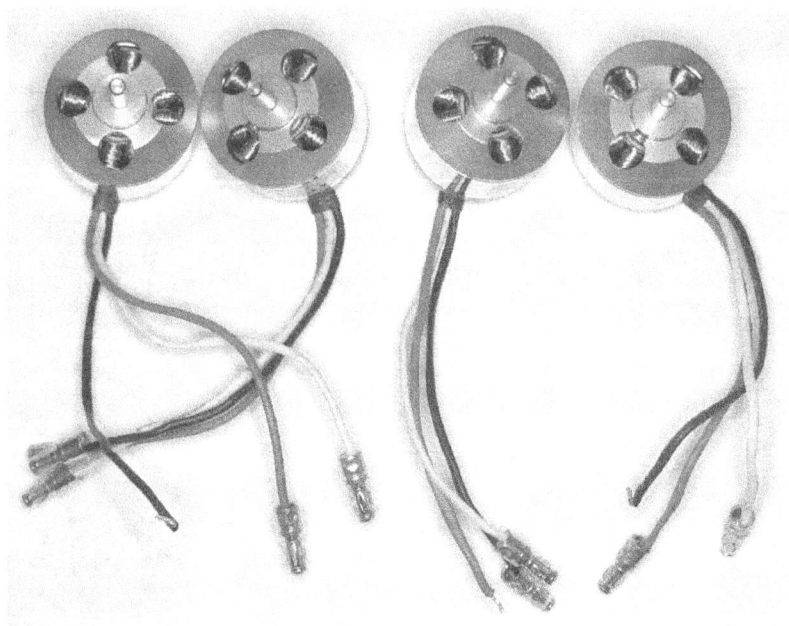

The ESC's (Electronic Speed Controllers) are rated at 30 Amps. They do not have any markings on them unless you remove the yellow heat shrink tubing. Underneath there is an Arduino processor, some level shifting logic and six power transistors.

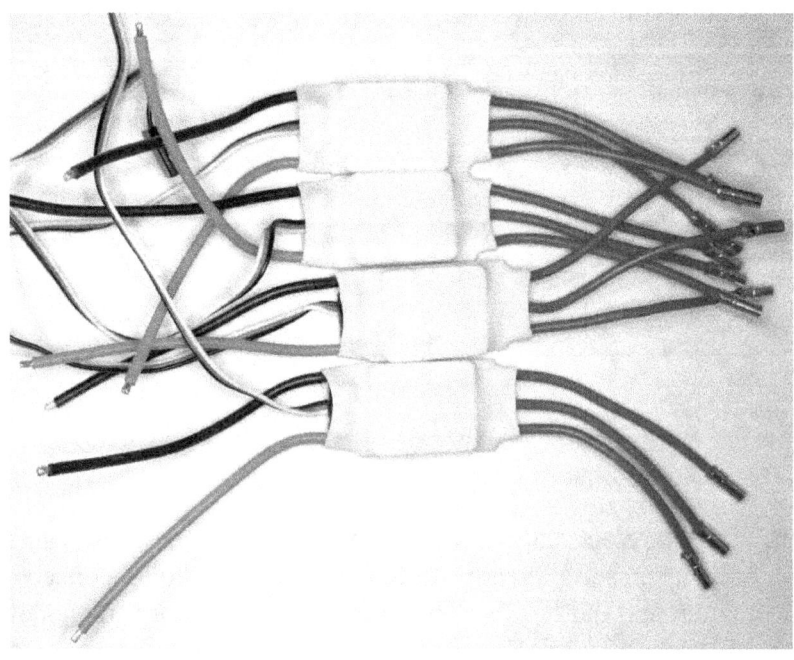

Then there are the propellers. The propellers are a set of 9443 propellers two are CW and two are CCW. You will not need the X shaped metal motor adapters that are shown with the propellers. The propeller number 9443 refers to 9.4 inches and 4.3 pitch.

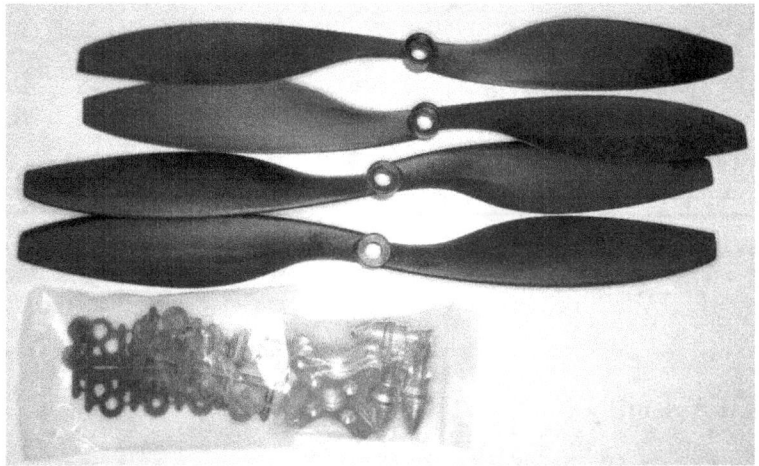

Last of all there are the miscellaneous small parts like heat shrink tubing and screws. My kit came with a bunch of Allen wrenches but not all kits will include them. Missing from the small parts picture is the Velcro battery strap that is used to hold the battery in place.

If you buy a complete quadcopter ARF kit or "Almost Ready to Fly" there are some of other things that you will likely get in that kit that did not come with my "partial" kit. Here is a list of those items.

- Set of 140 mm high landing skids.
- APM 2.8 ArduCopter flight controller with case.
- 6M GPS with Compass L5883.
- DJI GPS foldable antenna mount.
- Carbon fiber camera mount with damping balls.
- IMAX RC B3 Pro compact 3s balance battery charger.
- 11.1V 2200MAH 30C Li-Po battery.

F-450 Kit Assembly

The first part of assembly is to solder the bullet connectors on the three blue wires from the ESC's and the motors. The wires from the ESC's are large and solder fairly easily into the connectors. The wires from the motors are more difficult because they are much smaller. About three of my motor wires easily pulled out of the connectors after being soldered.

The solution is to bend the wires in an "L" shape so that there is more wire going into the bottom of the connector. Use an alligator clip to hold the connector in place. Heat it with the soldering iron and fill the end with solder. Then push the motor wire into the hot solder. After the solder cools give it a tug to make sure that it is solid.

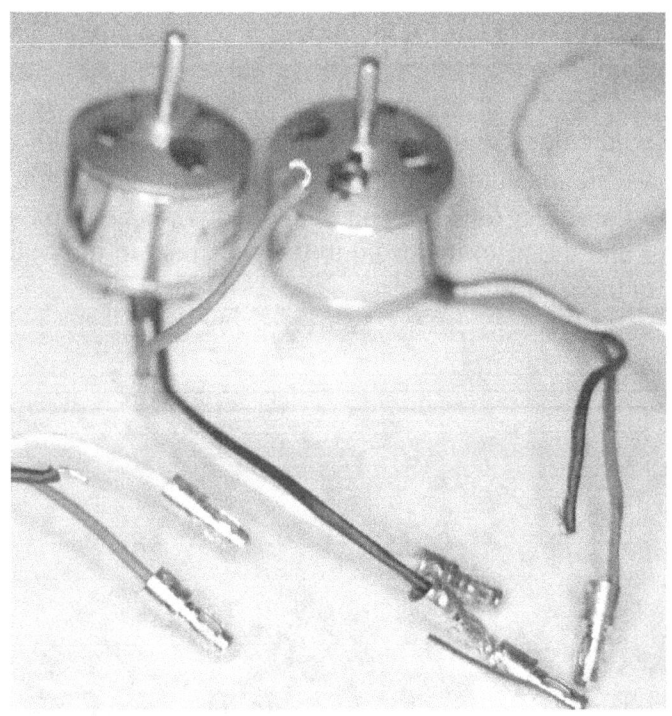

Once the connectors are soldered on, cut several 3/4 inch lengths of heat shrink tubing to go on the motor ends and one inch pieces to go on the ESC ends. Use a heat gun or soldering iron to shrink the tubing leaving the opening exposed so they can be plugged together. A hair dryer does not work that well as they are not hot enough.

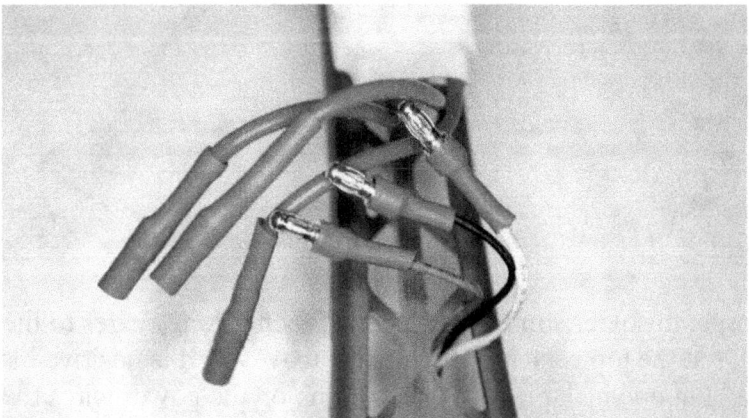

The next step is to solder the ESC's wires to the bottom of the frame. The center pieces of the frame are made out of printed circuit boards to make connecting the ESC's together a lot easier. There are power distribution runs that are built into the bottom of the frame.

First tin the gold colored pads on the frame and the wires with a coat of solder. Then while pushing down with the soldering iron slightly squish the leads into the solder onto the pads. The red wires go to the pads next to the + signs and the black wires go to the pads next to the – signs. This can be seen in the next picture.

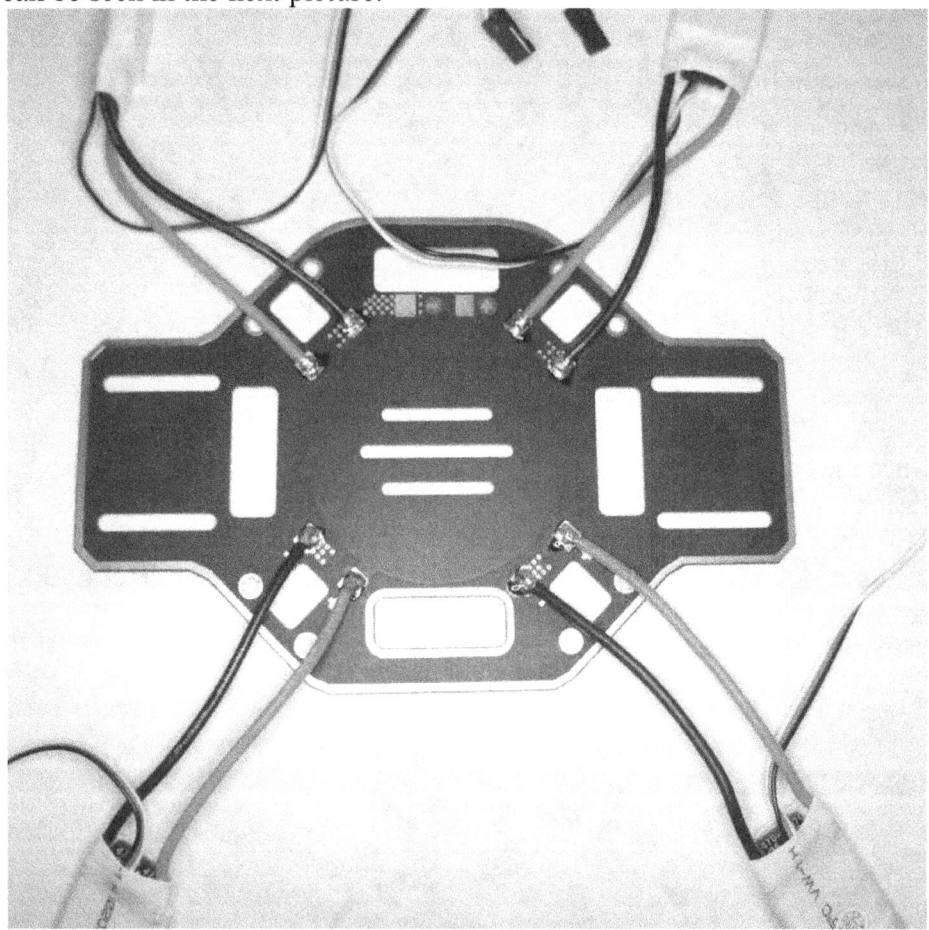

Don't forget to solder some short (about 5 inches long) wires to the power input pads at the top center in the picture above. Red is positive black is negative. Put one inch pieces of heat shrink on the power wires then solder a T connector on the other ends of the power input wires. This connector is hard to solder as there is no hole in the connector to hold the wire in place. The positive red wire goes to the pin that is turned sideways.

The next step is to attach the four legs to the bottom board. Pick a color such as white to indicate the front two legs and to differentiate them from the back two legs. The wires to the ESC's will fit in between the screws that fasten the legs to the body. The wires might be extra long and will need some bending to fit.

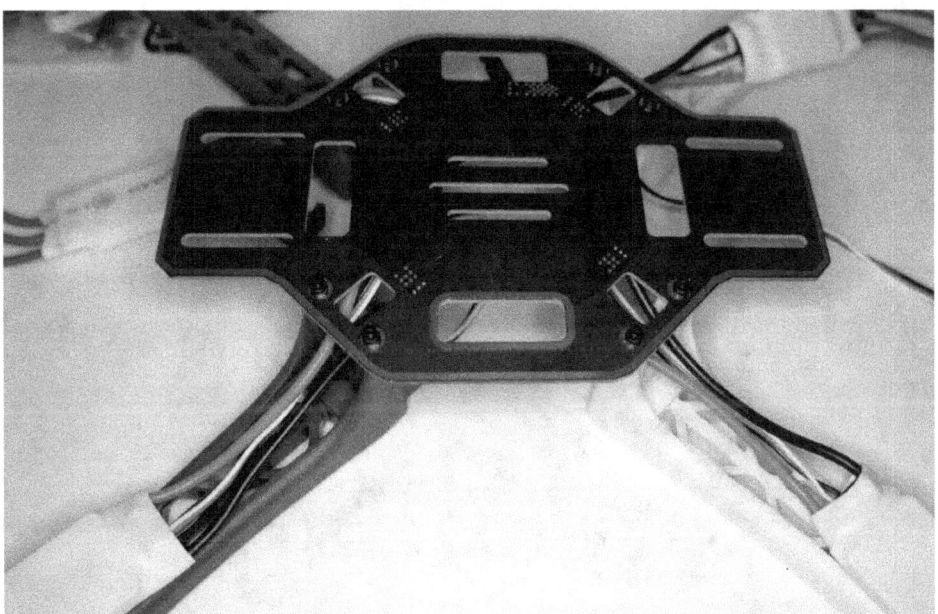

The ESC's are fastened to the legs with wire ties as close to the center of the frame as possible. My wires were extra long so they had to be pushed up inside the body first. Then add the motors with their wires headed toward the center of the quadcopter. Tighten the motor screws to about the same tension from side to side. Some people recommend using lock tight on the motor screws. Make sure they are tight.

Push the motor wires down through a hole in the frame and plug them into the ESC's. The order they are connected in is not critical at this point. A second wire tie will be needed to hold the wires tight to the frame. Once again my wires were a little longer than they needed to be.

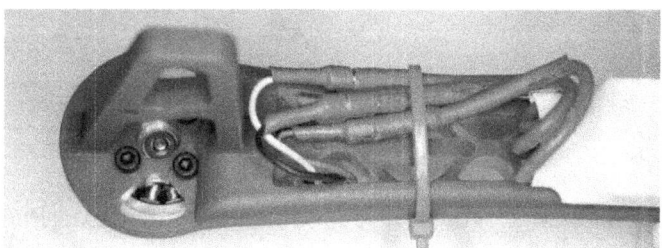

Next attach your flight controller to the top of the frame with some nylon spacers. Then screw the top piece to the rest of the drone. You can route the wires from the ESC's up through the top of the frame while you are assembling it.

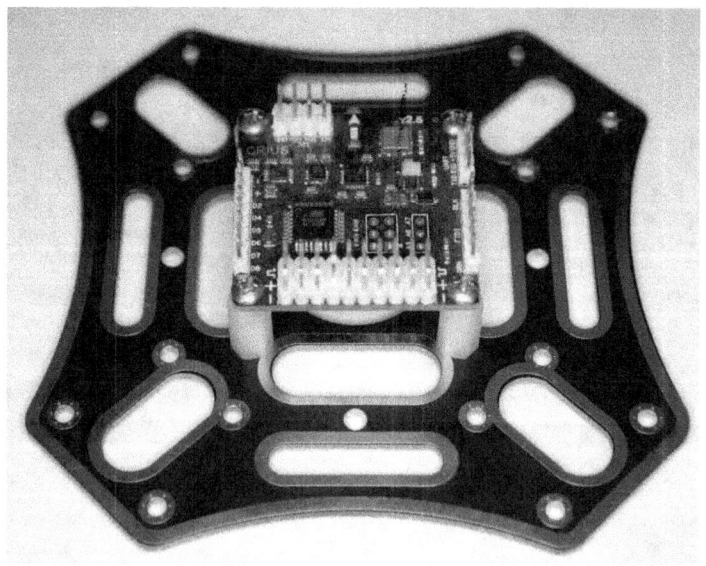

At this point the assembled drone should look something like what you see in this next picture. The front of the quadcopter is aimed towards you.

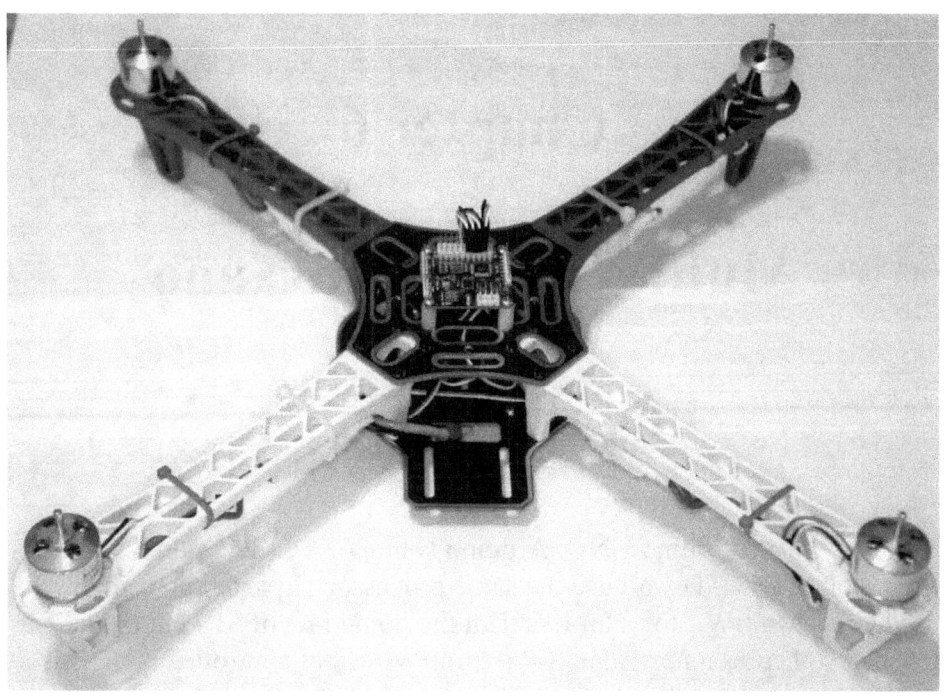

Next you will need to add the battery. It is held in place with Velcro on the bottom and a strap that goes around it. It should fit down through the hollow center (other than the ESC wires) of the drone.

At this point the assembly instructions will vary depending on the flight controller that you are using. The above picture shows a multiwii flight controller as the mounting holes fit one perfectly.

Chapter 6

MultiWii Uno and Nano

Flight Controller

You can use an Arduino UNO, Arduino Nano, or a Crius MultiWii as a flight controller. They all use the same processor. I picked up a slightly used Crius MultiWii for a lot less than the going rate of $19 but the ad was terrible, so here is a normal ad for a MultiWii flight controller.

CRIUS MultiWii Standard Edition Flight Controller MWC SE v2.6 Supported 2-axis G

$18.89   From China
Buy It Now
Free shipping

The people at Crius also make a MultiWii Pro. I is an Arduino Mega turned into a flight controller. It runs about $29 but has a lot of additional features. It should come with cable adapters as it has extra small connectors on it.

CRIUS All IN ONE PRO Flight Controller V2.0 Lastest Ver Pirate/MWC/ArduPlaneNG M
(122355424220)

Estimated delivery Wed, Aug 30 - Thu, Oct 12
Tracking number: 9400110895343028156393
This item has been shipped.

If you are using an Arduino Uno or Nano you will need a gyroscope device of some sort to help stabilize your flight. The MPU-6050 is a very popular gyroscope for a basic flight controller. These go for about $2 from China.

GY-521 3 Axis Accelerometer Gyroscope Module 6 DOF MPU-6050 Module for Arduino
(201004496054)

Estimated delivery **Wed, Mar 22 - Fri, May 05**
Tracking number: HK694443202PY

Another option for a gyroscope for use with an Arduino Uno or Nano is a 10DOF IMU or Inertia Measurement Unit. They run about $12 but offer altitude hold via an air pressure sensor.

10DOF IMU MPU6050 HMC5883L BMP085 Gyroscope Acceleration Compass Module Arduino
(181935368827)

Estimated delivery **Wed, Aug 16 - Wed, Sep 06**
Tracking number: LK316055393CN

This item has been shipped.

The wiring varies depending on the flight controller that you are using and the motors that you are using. The first wiring is to connect the ESC's to the controller. For the multiwii D10 goes to the front right side, D3 to the front left side. D9 goes to the back right side and D11 to the back left side. The white wire goes towards the center of the controller and the black (ground) wire goes towards the outer edge.

Later on, we will set the direction of rotation by swapping any two wires going from the ESC to the motor. The three pin connectors on the MultiWii are NOT in numerical order. The following diagram is for the MultiWii, Arduino Uno, Nano, and Mini. Basically, it is for all of the controllers that are based on the Atmel 328P Processor with the MultiWii software.

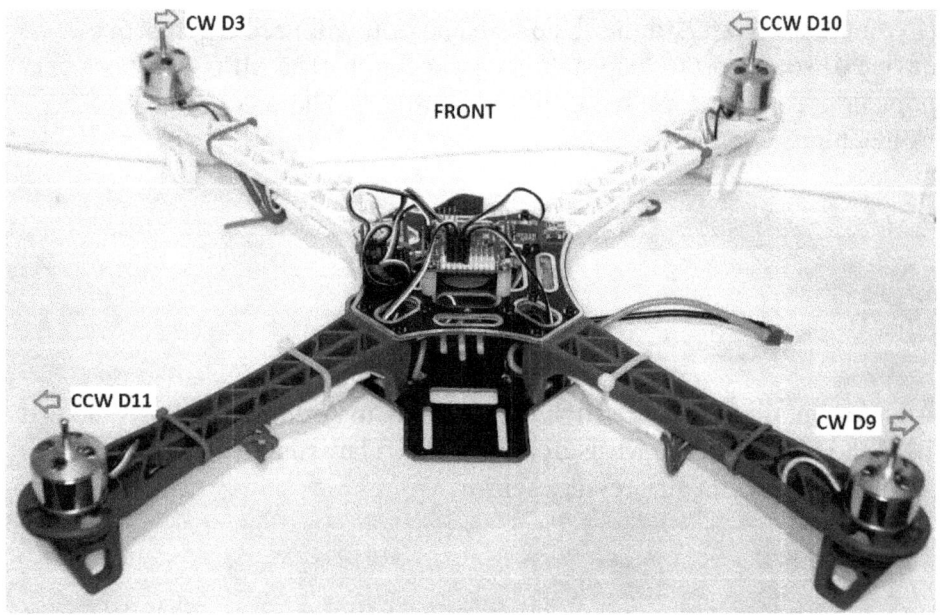

The next thing to connect up is the RC or Remote Control receiver. It is usually connected with two three pin connectors on the left side of the MultiWii. You can fasten the receiver next to the flight controller with a wire tie. On Uno or Nano setups the throttle goes to D2 and then the other three go to D4, D5 and D6.

The right most connector in the picture goes to the throttle or receiver channel 3, it also carries power to the receiver via the + and – terminals. The second, left, connector goes to the receiver channels 1, 2 and 4. The last three only need to connect to the innermost or "signal" terminals on the receiver. This reduced pin arrangement is done to make the multiwii controller smaller in size.

Here is a schematic diagram of how to connect the RC receiver to the MultiWii. The throttle on channel 3 uses all three pins because it is usually used to distribute power from the MultiWii to the remote control receiver.

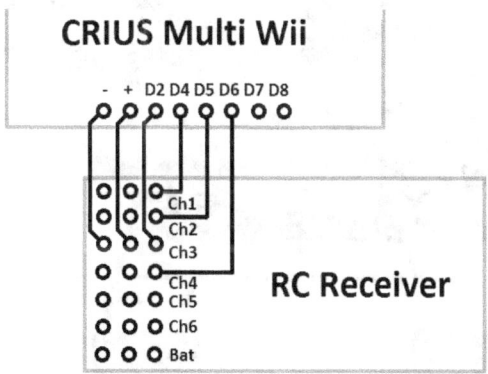

Up next is a large overall schematic diagram of the entire Multiwii or similarly controlled Quadcopter. Please note that the battery positive and negative connections are in the opposite order of the ESC positive and negative connectors.

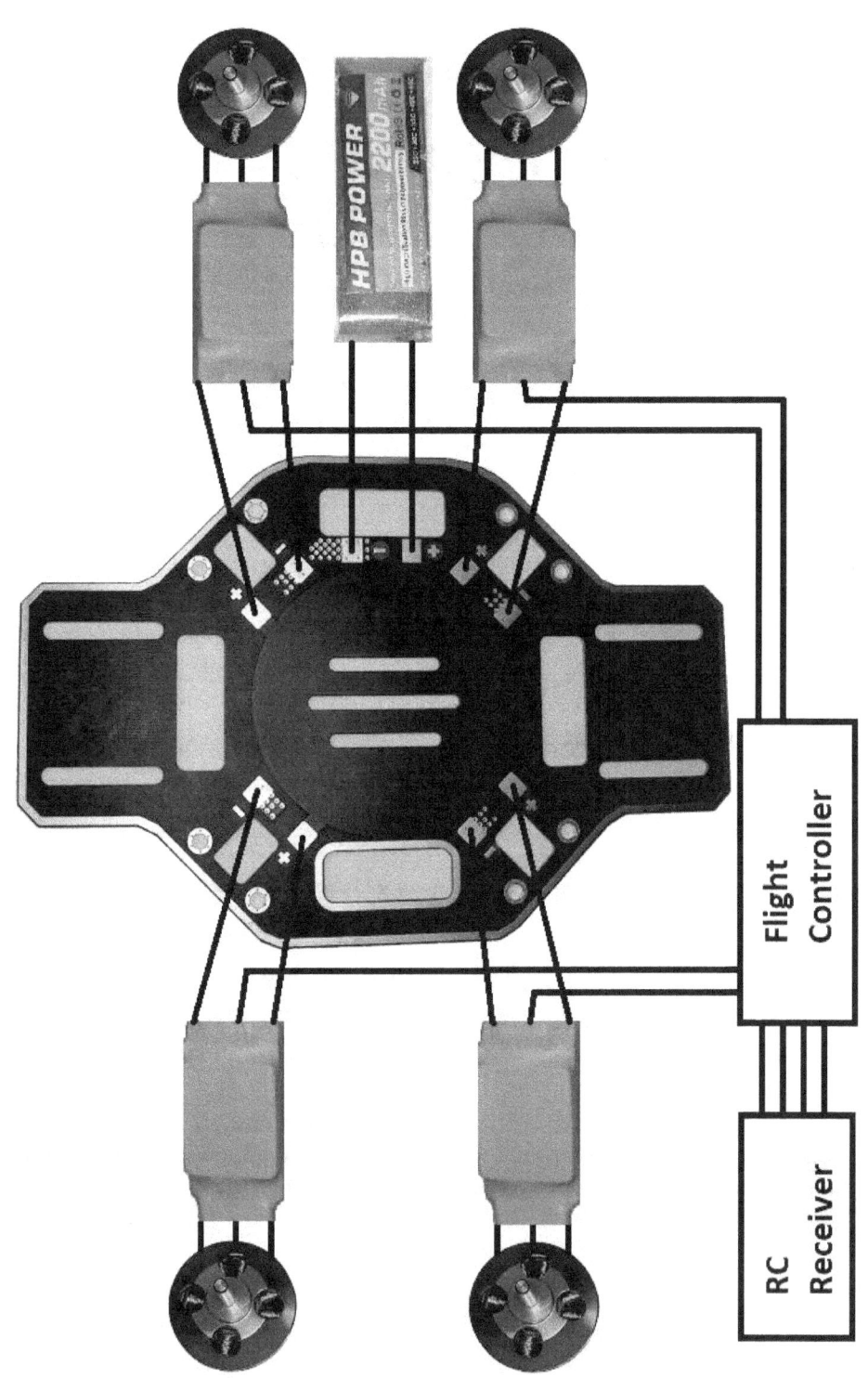

If you are using an Arduino Uno or Nano for a flight controller instead of a Multiwii you might need to make an adapter to mount the flight controller. The adapter is easily made out of a blank circuit board with holes already in it that are on .1 inch spacing. The board size is 2.1 inches by 2.3 inches. All the holes line up with the existing holes except for the one that is in the upper right corner.

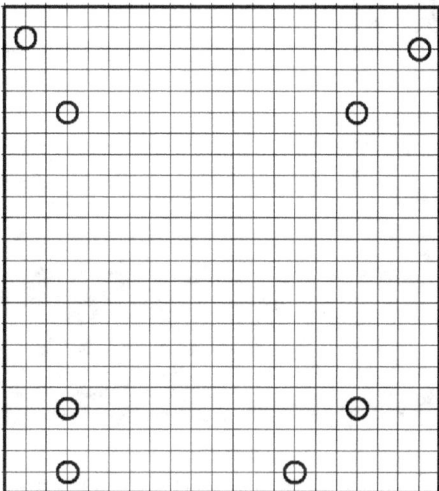

The adapter installs on the quadcopter after first installing the four spacers in the outer corners facing up. Then add four screws with either washers or nuts to space it up slightly to accommodate the screws that are holding the spacers. After those four screws go through the frame you can then add a second set of nuts below the frame.

The next picture shows the Arduino Uno adapter mounted on the top of a F-450 quadcopter frame. The RC receiver is fastened in with a wire tie to the left of the adapter board.

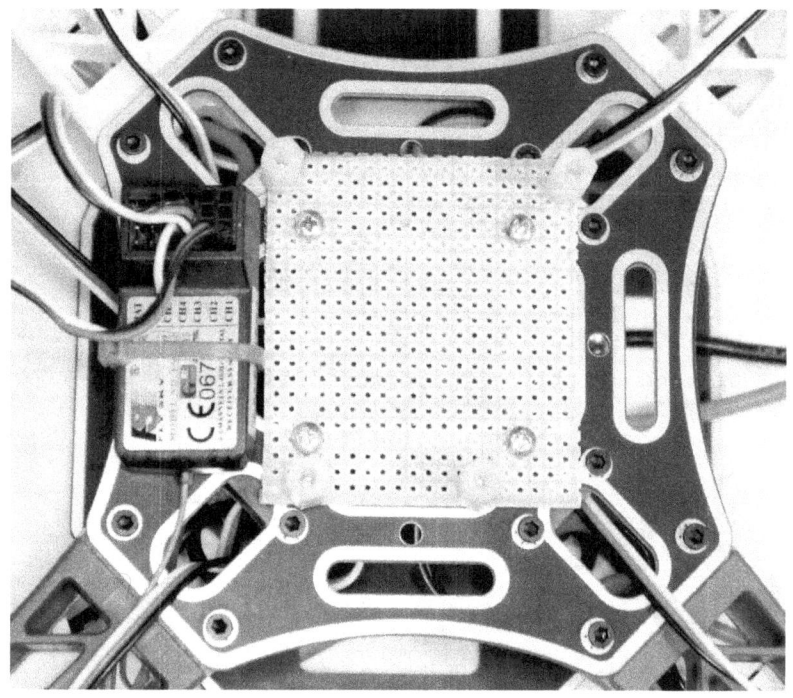

This adapter fits the Arduino Nano or UNO onto the Quad-Copter instead of the MultiWii. The Nano requires an expansion shield or homemade adapter, and the Uno requires the Sensor Shield or a homemade equivalent. They are both bigger and heavier than the MultiWii. This next picture shows the MultiWii with its USB interface adapter, the Nano with its expansion adapter, and the Uno with the sensor shield.

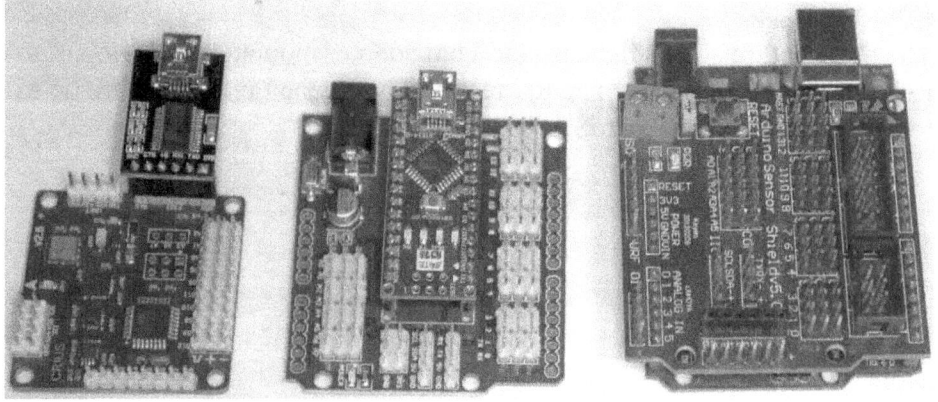

This next picture shows a Nano flight controller mounted on the quadcopter. The wiring gets a little messy when compared to the Crius Multiwii.

The use of the Arduino Nano or the Uno will require that you add a gyro like the GY-521. The GY-521 or equivalent must be oriented with the Y axis pointed forward and the X axis pointed to the left. It can be easily mounted on the left side of the Quad. The GY-521 does not have a compass so the quadcopter might tend to rotate during flight.

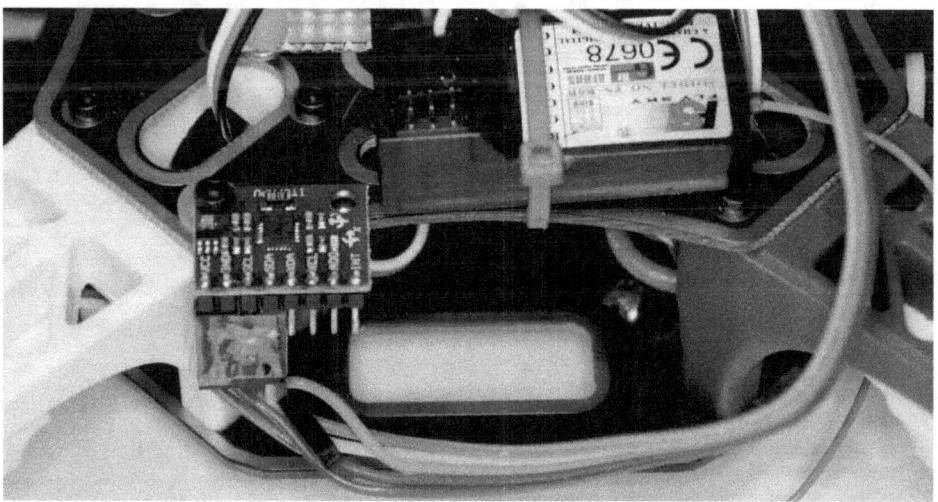

The GY-521 only requires that the first four wires to be connected to work. For both the Uno and Nano the SCL and SDA wires will need to be swapped between the GY-521 and the Flight controller. For the Nano the Vcc and Gnd wires will also have to be swapped.

The other end of the cable plugs into the 4-pin connector on the sensor shield that has the same pin names; they are just in a different order. I wrapped the pins in scotch tape to keep them in the correct order if they are accidentally unplugged.

MultiWii Setup

The downloaded MultiWii files come with the MultiWii software for the flight controller as well as several versions of software to run on your computer to communicate with the flight controller. This is necessary to get the MultiWii configured properly and calibrated before flying. The following picture shows the Windows 32 bit executable.

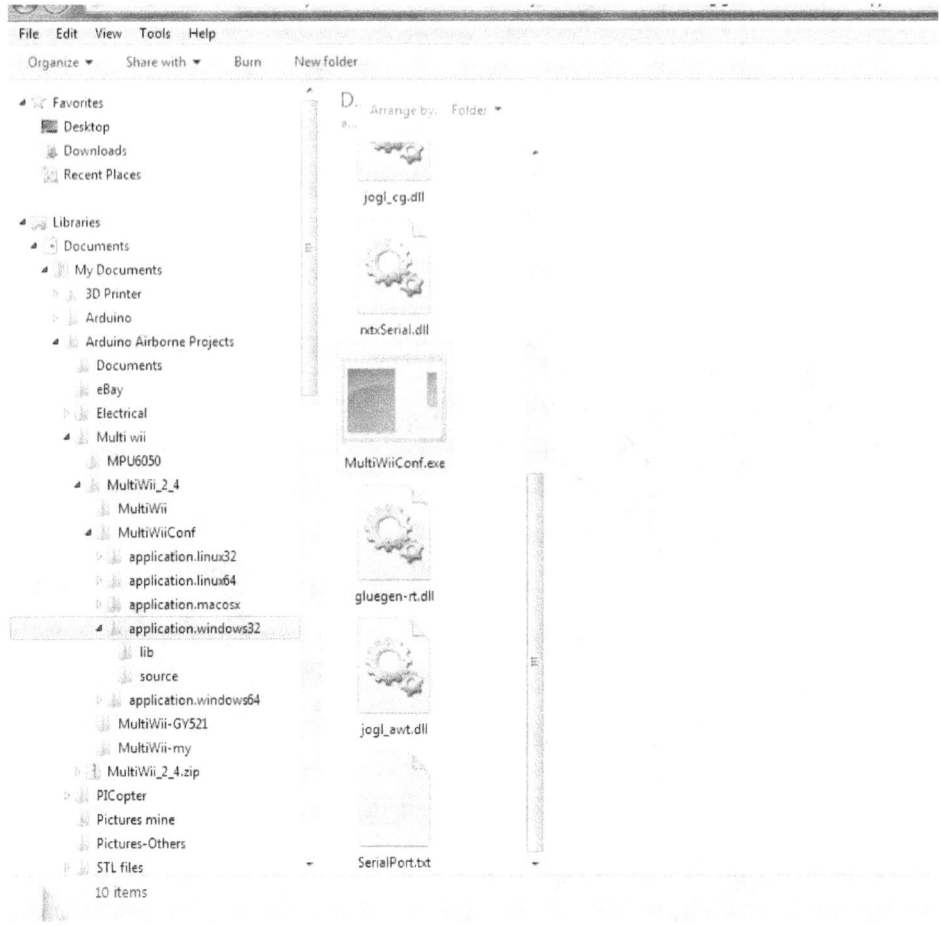

To configure the MultiWii flight controller you need to install and load the Arduino software, sometimes called the Arduino IDE. It should have a round blue icon with a sideways 8 in the middle. When you first plug in the flight controller you will have to tell Windows where to find the driver, it is in the Arduino directory. Next select and open the MultiWii ino file for the flight controller. Then you need to find the config.h tab. It is a tab that is all the way to the right, then a drop-down list, then scroll all the way to the bottom of that list. You can see this pattern in the next picture.

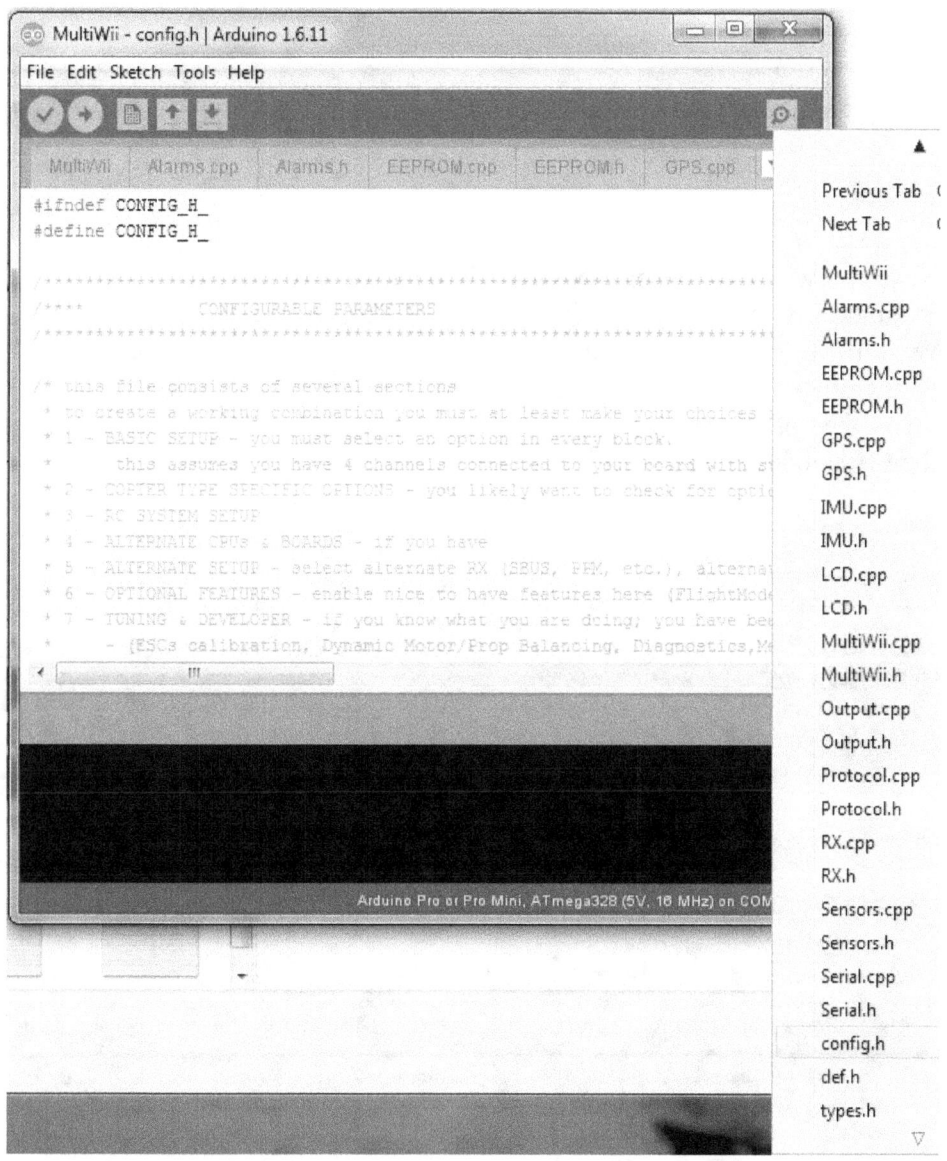

Once you get into the config.h file there are some things you will need to change in it. To enable a selection, you will need to remove the double backslash in front of the line. First of all, you need to select what your flying configuration is. Usually that is the "Quad X" configuration. So you would remove the"//" that is in front of "QUADX" configuration to select that.

```
/*************************    The type of multicopter    *********************/
//#define GIMBAL
//#define BI
//#define TRI
//#define QUADP
#define QUADX
//#define Y4
//#define Y6
//#define HEX6
//#define HEX6X
//#define HEX6H    // New Model
```

Next you will need to select your flight controller. For the CRIUS Standard Edition enable the line that says "#define CRIUS_SE_v2_0" for version 2 or newer. There are support files for a lot of other controllers. If your flight controller is not fully supported then you have to individually select the options that you are using such as the GY-521 gyroscope with the MPU6050.

```
//#define CITRUSv2_1          // CITRUS from qcrc.ca
//#define CHERRY6DOFv1_0
//#define DROTEK_10DOF        // Drotek 10DOF with ITG3200, BMA180, HMC5883, BMP085, w or w/
//#define DROTEK_10DOF_MS     // Drotek 10DOF with ITG3200, BMA180, HMC5883, MS5611, LLC
//#define DROTEK_6DOFv2       // Drotek 6DOF v2
//#define DROTEK_6DOF_MPU     // Drotek 6DOF with MPU6050
//#define DROTEK_10DOF_MPU//
//#define MONGOOSE1_0         // mongoose 1.0    http://store.ckdevices.com/
//#define CRIUS_LITE          // Crius MultiWii Lite
//#define CRIUS_SE            // Crius MultiWii SE
//#define CRIUS_SE_v2_0       // Crius MultiWii SE 2.0 with MPU6050, HMC5883 and BMP085
//#define OPENLRSv2MULTI      // OpenLRS v2 Multi Rc Receiver board including ITG3205 and AD
//#define BOARD_PROTO_1       // with MPU6050 + HMC5883L + MS baro
//#define BOARD_PROTO_2       // with MPU6050 + slave  MAG3110 + MS baro
//#define GY_80               // Chinese 10 DOF with  L3G4200D ADXL345 HMC5883L BMP085, LLC
//#define GY_85               // Chinese 9 DOF with   ITG3205 ADXL345 HMC5883L LLC
//#define GY_86               // Chinese 10 DOF with  MPU6050 HMC5883L MS5611, LLC
//#define GY_88 // Chinese 10 DOF with MPU6050 HMC5883L BMP085, LLC
#define GY_521               // Chinese 6 DOF with  MPU6050, LLC
//#define INNOVWORKS_10DOF    // with ITG3200, BMA180, HMC5883, BMP085 available here http:
//#define INNOVWORKS_6DOF     // with ITG3200, BMA180 available here http://www.diymulticopt
//#define MultiWiiMega        // MEGA + MPU6050+HMC5883L+MS5611 available here http://www.di
//#define PROTO_DIY           // 10DOF mega board
//#define IOI_MINI_MULTIWII// www.bambucopter.com
//#define Bobs_6DOF_V1        // BobsQuads 6DOF V1 with ITG3200 & BMA180
//#define Bobs_9DOF_V1        // BobsQuads 9DOF V1 with ITG3200, BMA180 & HMC5883L
```

Then you will need to select what the Arduino IDE software is supposed to upload your flight controller software into. At the top select "Tools" and "Board". The normal setting is Arduino Pro or Pro Mini and then set the processor to: "Atmega328 5V 16MHZ". You will also have to select the USB Port that it is connected to.

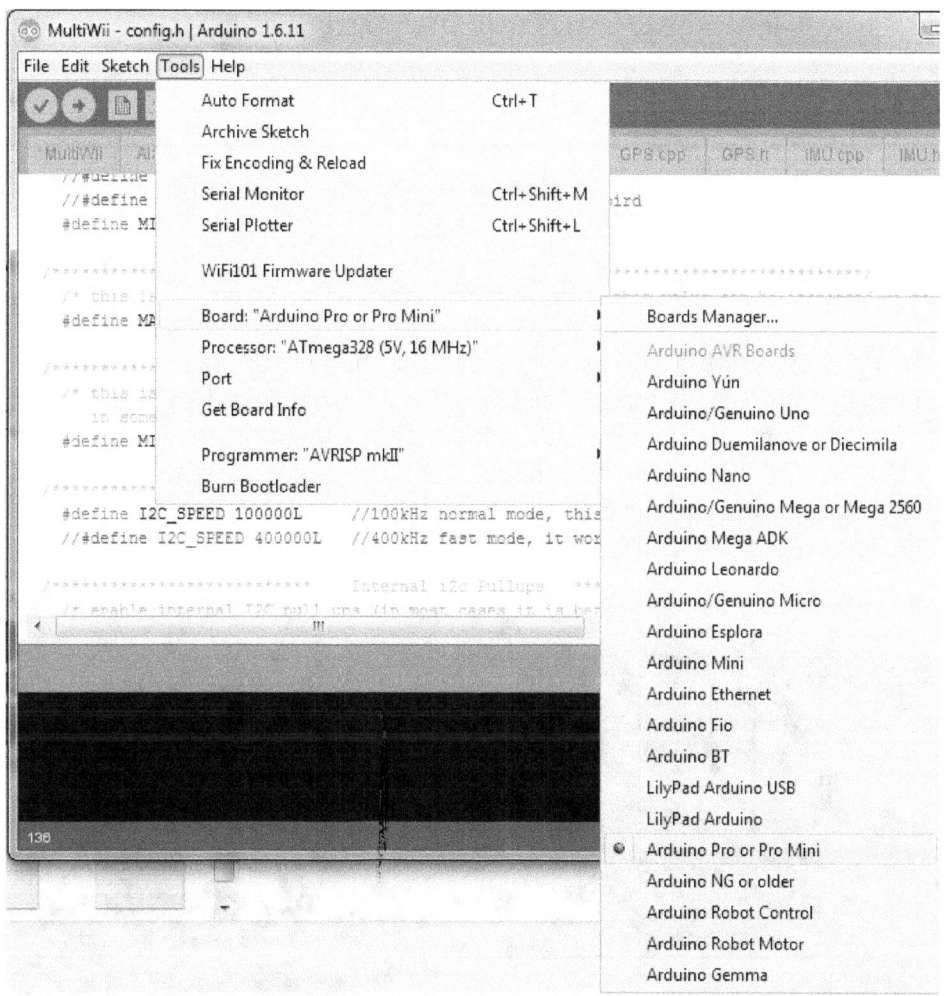

Once you have made all of your changes to config.h and selected your flight controller, board, processor and port you will then need to upload the revised flight controller software to your MultiWii or Arduino based flight controller. Usually that is done by selecting "Sketch" and "Upload". That upload location has moved in some versions of the Arduino IDE software.

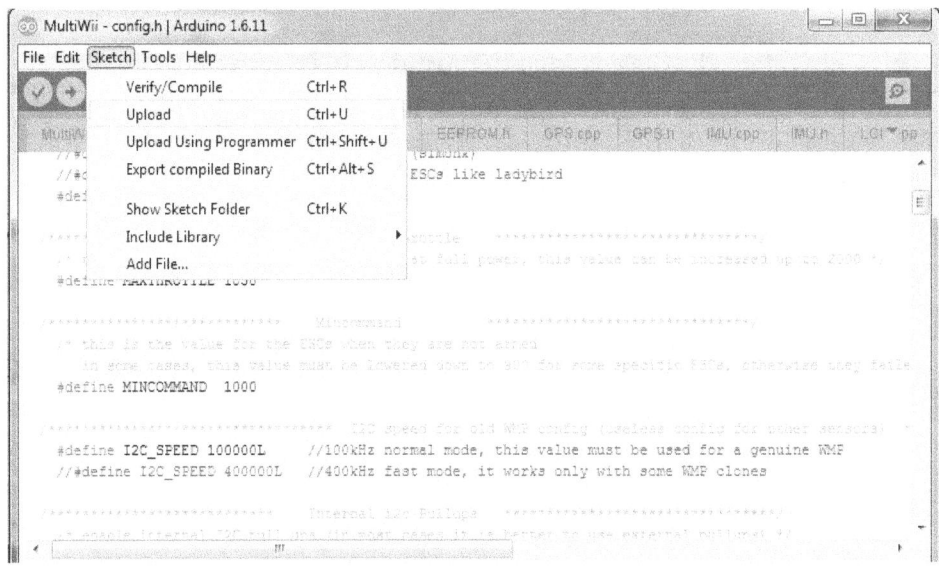

Then go to your MultiWii extracted files and find the application program directory to fit your computer such as "Windows64" and then find "MultiWiiConf.exe" and run that. Select your communications port in the "PORT COM" box. That box should then turn green. Next select "Start" and the monitoring of the accelerometer, gyroscope, altimeter and magnetometer should start.

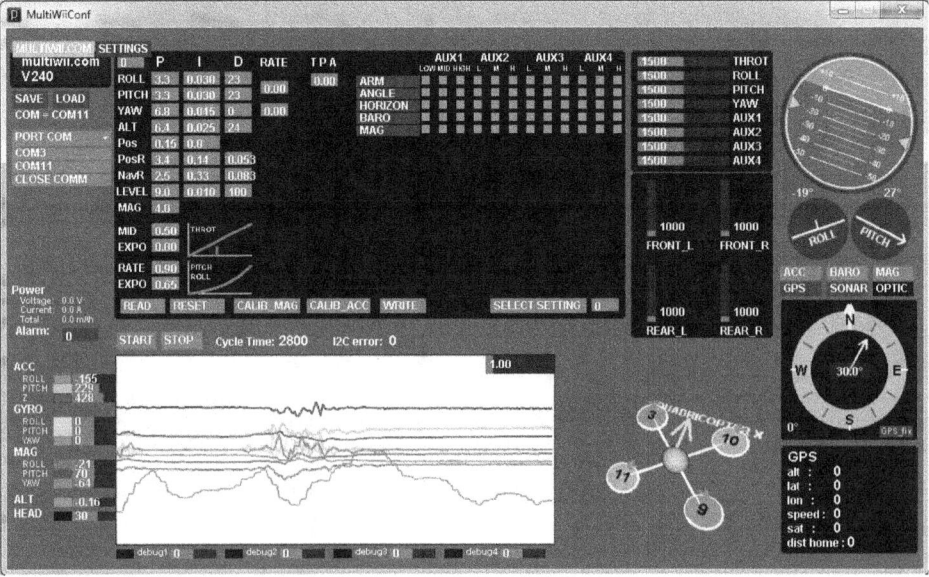

If you get lots of I2C errors then your flight controller is not configured correctly or it is not working properly.

50

In this screen you can calibrate the magnetometer of the MultiWii. To calibrate the magnetometer, you will need to position the quad nose down, nose up, on the right side, and the left side, as well as rotate it 360 degrees. A blue light blinks during the calibration procedure to let you know it is still working on it. To calibrate the accelerometer, it needs to be still and flat. These must be calibrated because the MultiWii will not "arm" the motors if the quadcopter is not "flat".

If your flight controller is connected and working you can also trim your Roll, Pitch, and Yaw to as close to "1500" as possible. The Throttle should trim close to "1000" when it is all the way down.

If you have not manually calibrated your ESC's the multiwii configuration program has a modification that can only be used to calibrate the ESC's. Near the bottom of config.h there is a line where you can remove the // to #define ESC_CALIB_CANNOT_FLY. Then upload the program. Disconnect and power up the drone with the battery until all the beeping stops. Then power it down, reconnect, put the // back and re-upload the multiwii program.

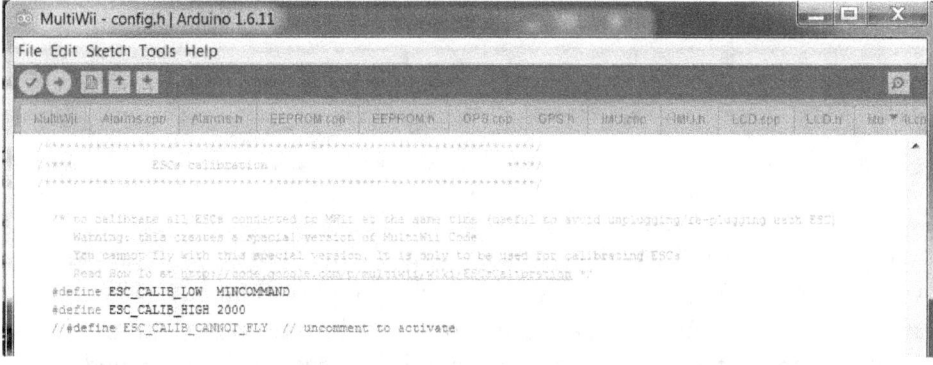

At this point you can push the throttle down and to the right to "arm" the motors. There should be no battery connected at this point. If everything is correct the "Arm" should turn red and the throttle should vary the front and rear motor speed bar graphs. Click "write" to save your settings.

Next, we need to make sure the motors rotate in the correct directions. For this test there should be no propellers attached! Disconnect the USB cable and connect the battery. Pull the throttle back and all the way to the right to arm the motors. Once in a while they may not all start up. By starting and stopping the motors determine the direction that they are spinning one at a time. Write it down. Disconnect the battery and swap any two wires on any motor that is spinning in the wrong direction. Then reconnect the battery and check the direction of rotation once again.

Next is to add the propellers. Even this step is a bit tricky. The top, rounded, edge of the blades needs to match the motors direction of rotation. There are round spacers that need to be inserted into each blade to match the size of the shaft they are going on. In some cases, the insert needs to be one size larger to allow the collet to tighten on the motor shaft. Insert a small Allen wrench into the bullet to tighten the blade down. Basically, you have the collet, the chuck (enlarged side up), the blade then the bullet.

In this next picture the top propeller is for clockwise rotation and the bottom propeller is for counter clockwise rotation. The leading edge of the propeller is thicker and the trailing edge is thinner.

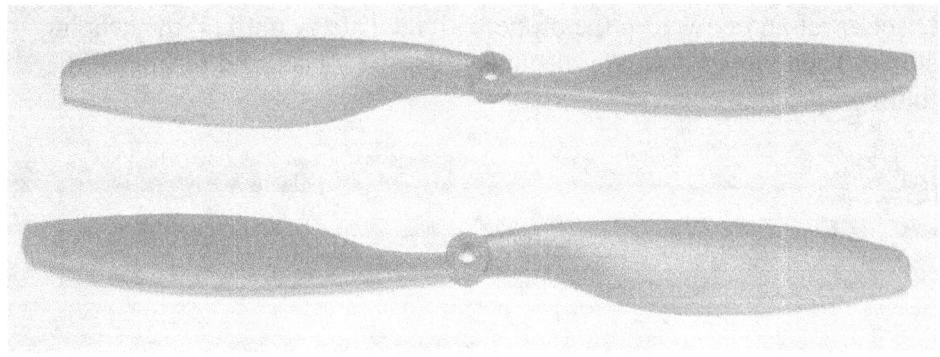

You are now ready for your first test flight. Some people recommend tying it to something for the first flight so it does not fly away. If you have done everything correctly that should not be a problem. In my first flight it flipped over and broke two propeller blades. That happened because I had the GY-521 on the wrong side of the quad.

Chapter 7

APM

Flight Controller

Another flight controller option is the ArduCopter, ArduPilot or "APM". It is based on the Arduino Mega Processor. As a result, it is faster and has many more Input and Output pins. This flight controller runs about $29 each. If you can get a flight controller that comes with the four sets of three pin cables that connect it up to the remote control receiver it will save you the trouble of making up your own cables.

APM 2.6 Flight Controller Board For Multicopter ARDUPILOT MEGA 2.6 Version H5
(111774405086)

Estimated delivery **Wed, Apr 19 - Tue, May 09**

A GPS can be added the ArduCopter to be able to send it on "Missions". The GPS runs about $12.

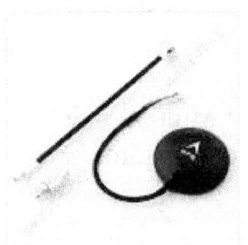

UBLOX NEO-6M GPS Module with Shell & Mount Support for APM 2.6 2.8 se
(161332562575)

Estimated delivery **Mon, May 22 - Tue, Jun 27**

In order to monitor the state of the battery you will need a power monitor adapter like this that sells for about $5.

The APM is an Arduino with a much more "mega" powerful processor. This next picture compares the MultiWii on the left and the APM board on the right.

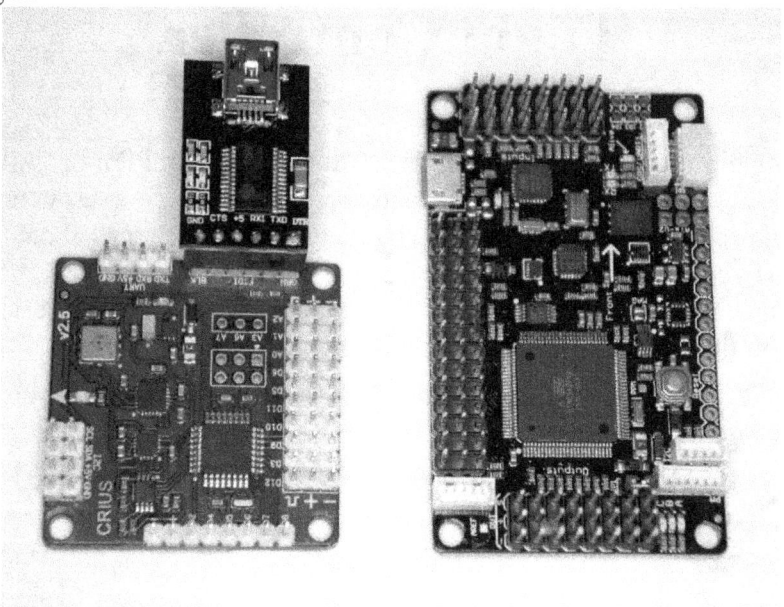

The APM flight controller is very different from the MultiWii. Primarily it has eight "inputs" and eight "outputs". It also has a jack for a GPS device, a jack for a power monitor device, and a jack for an on-screen display and/or for radio telemetry.

The motors are wired up in a different sequence from the MultiWii. The APM uses outputs 1 to 4 for a four motor device. The next picture shows an APM with the GPS option mounted on a F450 quadcopter.

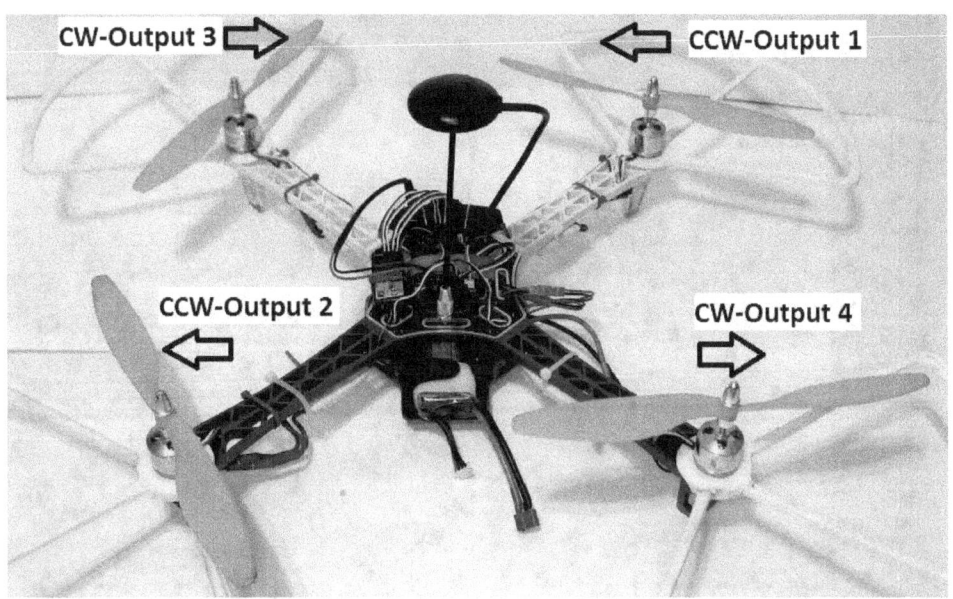

The software setup is also very different. It uses a wizard to guide you through the process. It also uploads the firmware into the flight controller once you have chosen your configuration. The software is called "Mission Planner".

Mission Planner also calibrates everything to your remote control as far as highest and lowest settings as well as the midpoints.

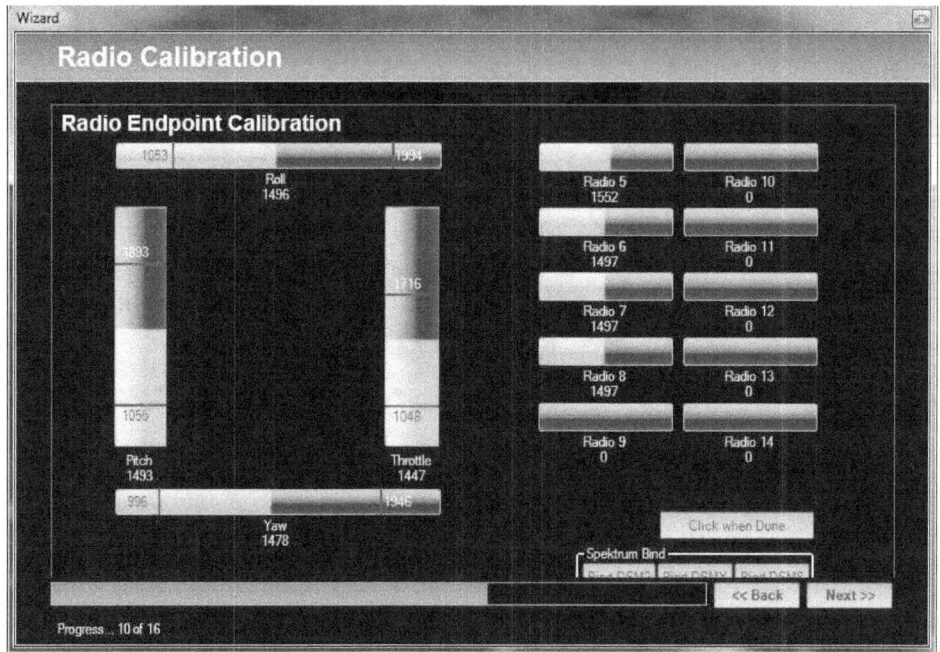

Once you are done with calibration and the firmware is updated then the GPS screen should come up with your exact location. This gives the quad the ability to return to the exact location where it took off as well as the ability to navigate to another location using GPS.

The next step is to calibrate the ESC's. This is done without any propellers installed. Turn on the remote control and turn up the throttle to maximum. Then connect the battery to the quadcopter. The APM should make a number of beeps saying it is ready to be powered off and back on. Disconnect and reconnect power. After the beeps turn the throttle down to zero and it should beep again saying that the ESC's are calibrated.

This next picture is a close up of the APM flight controller mounted on the F450 frame. Use the soft foam tape with peal-and-stick surfaces on both sides. A strap or wire tie can provide more safety in case of a crash. You can also see the dash cam on the front and the battery in the back in this picture.

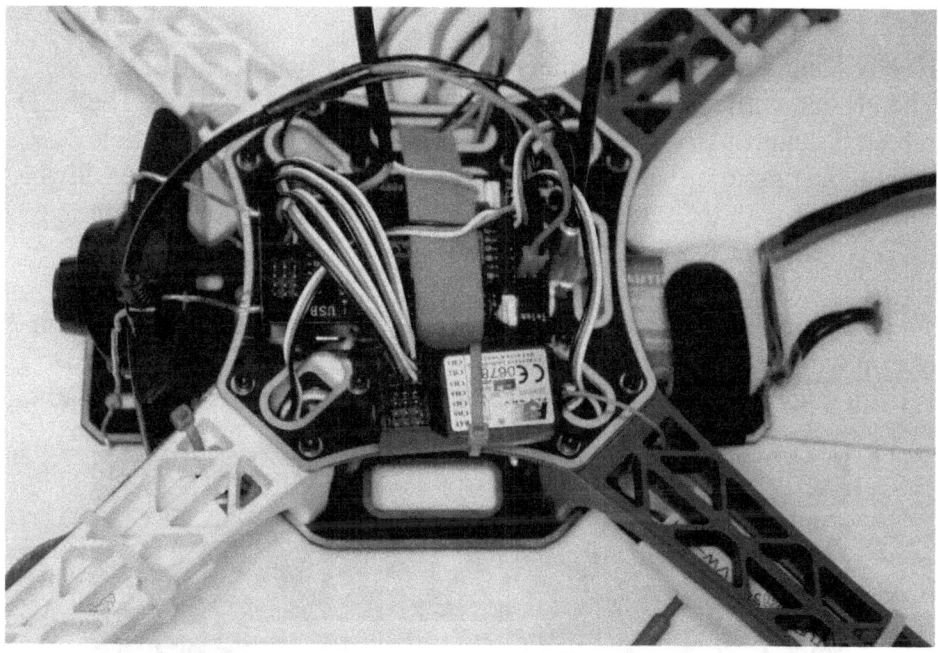

Chapter 8

Hobby King KK

Flight Controllers

Hobby King KK flight controllers come with and without a LCD Screen. The version with the LCD comes in "normal" and "mini" sizes. The LCD screen is a huge asset when it comes to changing settings in the field without having to use a laptop computer and some sort of interface device. The version with the screen is the KK2.1.xx and the version without the LCD is version KK5.5.xx

These controllers also properly support a hexacopter. That was the main reason why I decided to buy a few of them and try them out. The LCD version quickly became my favorite as it works great right out of the box.

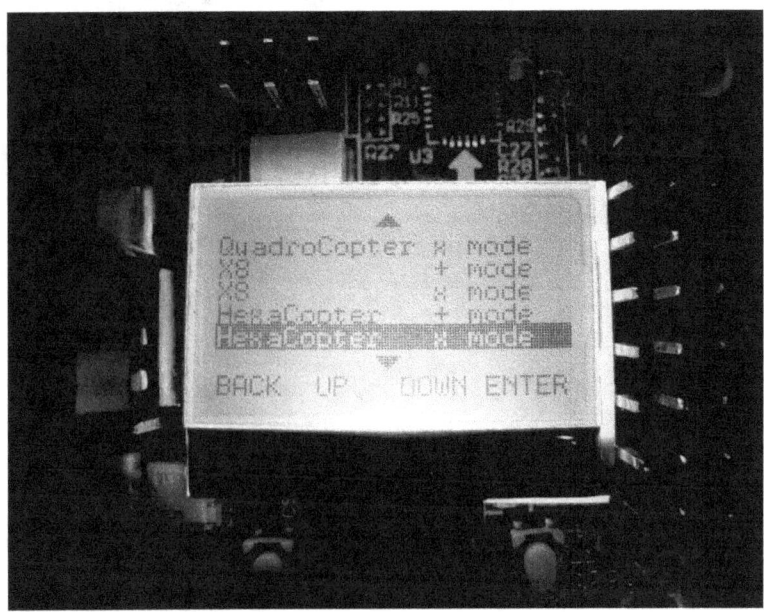

The LCD makes it easy to do a gyro calibration. That is often necessary as it does not seem to remember the last calibration. Changing the mode and configuration can easily be done with the LCD.

The older KK5.5 version has some tricky calibration procedures. For instance; to calibrate the ESCs you turn the left most "Yaw" potentiometer all the way down then power it on. There will be more on that later. The first problem is that the KK5 controller comes configured as a quad in the "+" arrangement. To support that setup, you will need to make an adapter like in the next picture to rotate it 45 degrees to fit a F-450 frame.

The other option is to reprogram the KK5 controller. This is not easy either. To reprogram the KK5 you will need a USB adapter and a 10 pin to 6 pin cable. No one seems to be clear as to how to plug the 6-pin adapter into the controller! After some testing I determined that it is with the cable passing over the pitch sensor. I will show you how to do that in the next picture.

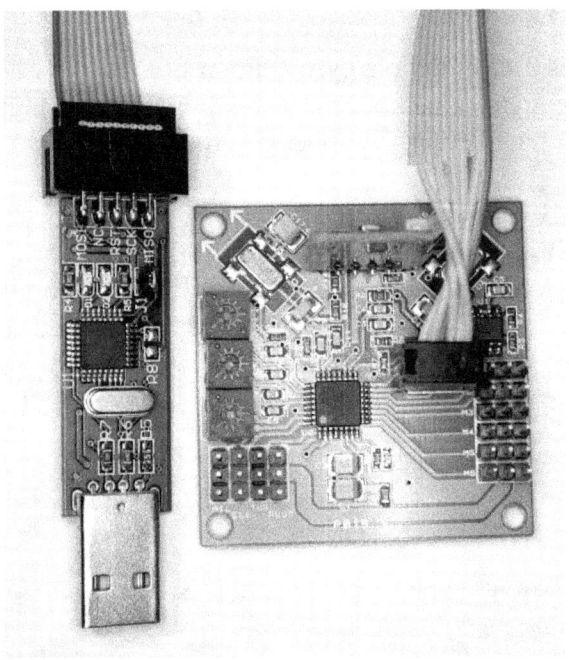

After you plug it in you will need to set up the USBasp diver. It can be found at the Fischl.de web site. You will ed to unzip the file to a location then tell Windows, Control Panel, Deice Manager, where to find it.

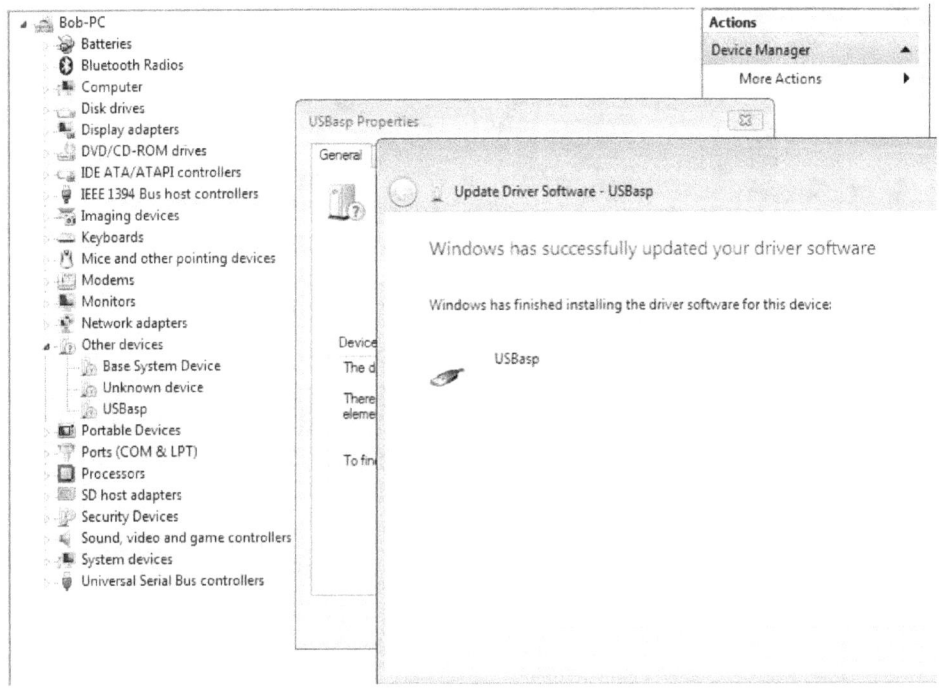

Next you will need to go to lazyzero.de/en to get the "KK Multicopter Flashtool". Unzip the file and run it. The Flashtool requires Java to run. You should get something like what you see in the next picture.

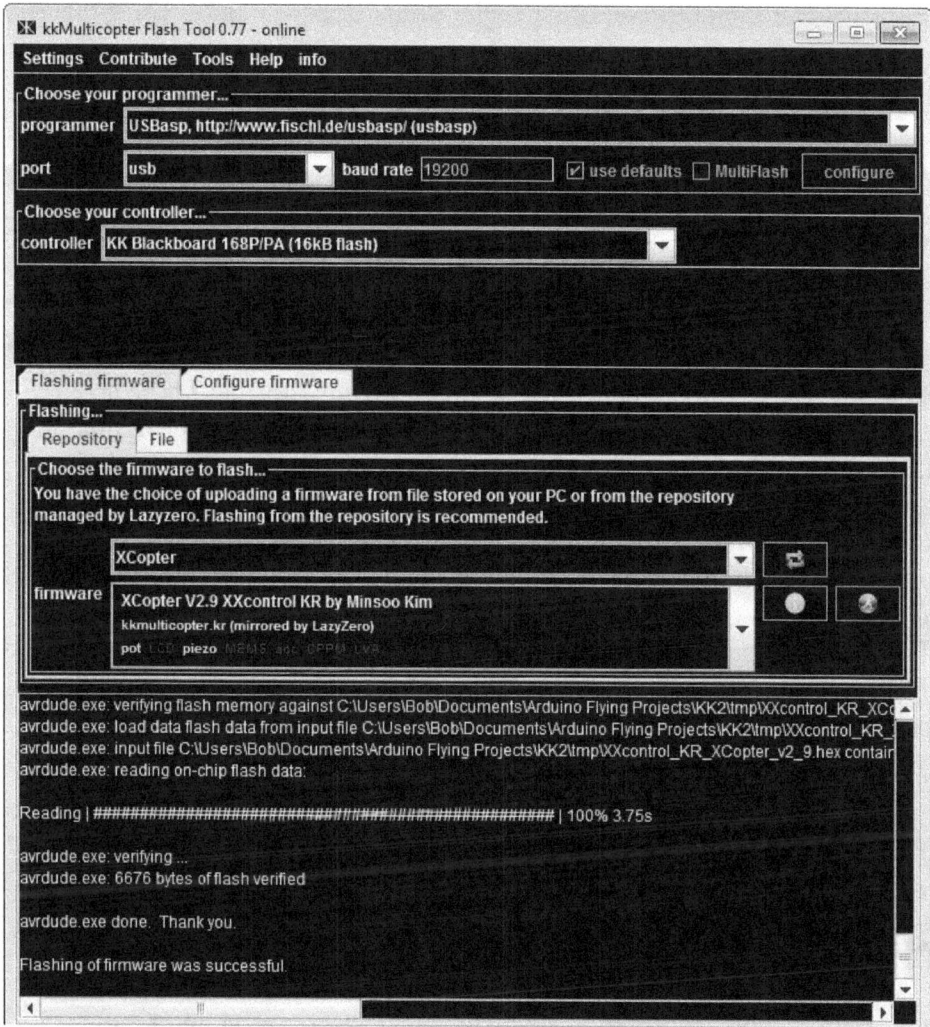

XCopter Version 2.9 worked the best for me for an "X" arrangement quadcopter flight controller.

Chapter 9

The ZMR 250

Racing Quad

Another type of quadcopter is the smaller "racing" quads. These are generally 220 to 250mm in wingspan. They are much smaller lighter and faster. There are quad races with prizes of up to a million dollars going to the winner. Racing quads require FPV video operation. They race through glow in the dark hoops and many other neat obstacles. You can only do that at high speeds with FPV goggles.

The frame is about $20 but it did not come with a power distribution board. That is because some builders use a quad ESC that does not require any power distribution board.

Glass fiber ZMR250 4 Axis 250 MM FPV Quadcopter Mini H Quad Frame for QAV250
(111953466377)

Delivered on **Wed, Jun 07**
Tracking number: **LK257911265CN**

The racing quad uses smaller, faster motors. These are HobbyFans HM2204U KV2300 motors. They are shorter and much lighter than the other motors. They cost a little over $4 each or $17 for 4 of them.

4x HM2204U 2300KV Brushless Motor CW & CCW for FPV 250 Mini Quadcopter QAV250
(201905086652)

Delivered on **Sep 07, 2017**
Tracking number: **LK334660941CN**

This next picture compares the two motor sizes. The left motor is a RS2212 - 920KV and the right motor is a HM2204 - 2300KV. The right motor is about 1/2 the height of the left motor.

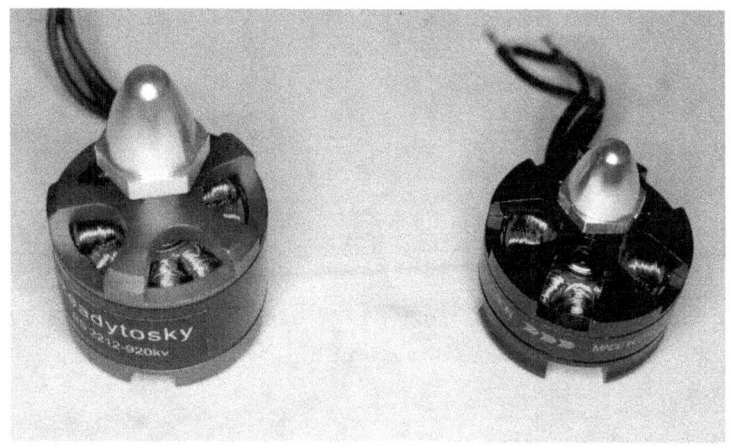

These smaller quads require smaller propellers. Usually five-inch propellers are used like a 5040 or a 5030. Sometimes, to get more lift, triple and quad blades are also used. These propellers were about $6 for eight of them on eBay.

8pcs Gemfan 5030 ABS 3-blade Propeller for Mini FPV Quadcopter QAV250 QAV280 OR
(152106251790)

Delivered on Sat, Jun 03
Tracking number: LK257395544CN

For more aggressive propellers you might want to try these 5 inch 50 degree tilt blades. They were only $4 for four of them. The greater tilt gives more lift but uses more battery power.

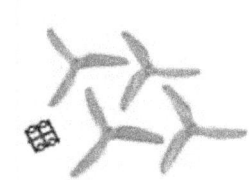

LUM7023 Lumenier 5x5x3 Butter Cutter Propeller (4) (Tangerine)
(122491102781)

Delivered on Sep 08, 2017
Tracking number: 9400110897864000900875

Next, to save room you can use a 4in1 ESC. It is limited to 12 amps but is a lot smaller and lighter than using four normal ESC's. In fact it was too small for the ZMR 250's mounting holes! This 4in1 ESC comes with no wires so you will need to add 12 about 3.5 inch wires to the bullet connectors for the motors.

BS412 Blheli_s 2-4S 4in1 ESC 4x12A Support Dshot/Multishot/Oneshot Quadcopter
(262991364711)

Delivered on Sep 11, 2017

Tracking number: LK334453595CN

This item has been shipped

This next picture compares the size of the ESC's. As you can see the 4in1 ESC is smaller than a typical 20 or 30 amp ESC.

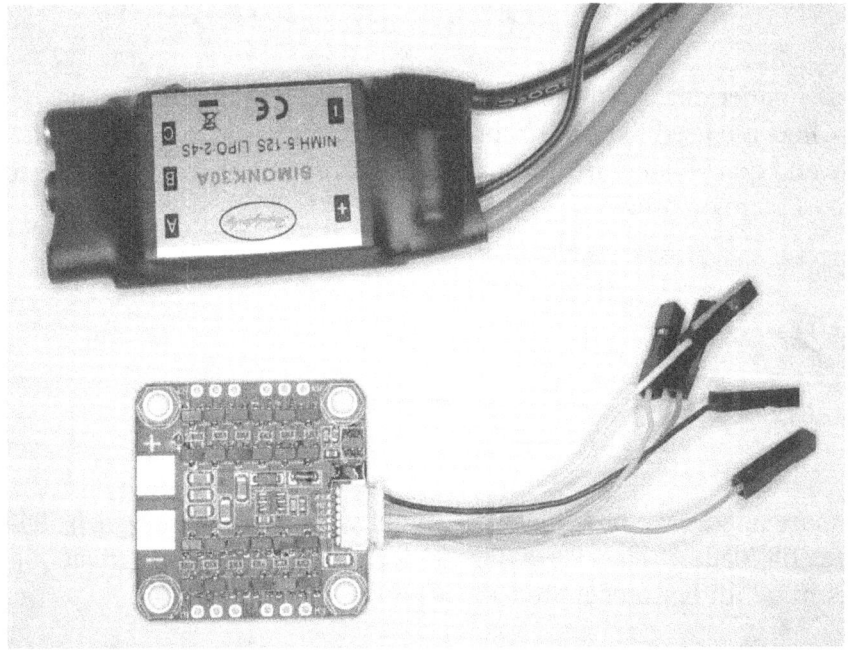

Also for the racing quad you might want to consider using a smaller battery. That reduces the flight time but it also takes up less room. This battery was only a little over $13. The smaller battery can fit on the lower level of the quad.

Lion Power Lipo Battery 11.1V 1500Mah 40C MAX 60C 3S For RC Car Helicopter
(111470192309)

🛈 Estimated delivery **Mon, Sep 18 - Tue, Sep 26**
Tracking number: **74899991206461354877**

This is what the ZMR 250 frame kit looks like. It came with two bags of screws and a one page instruction sheet.

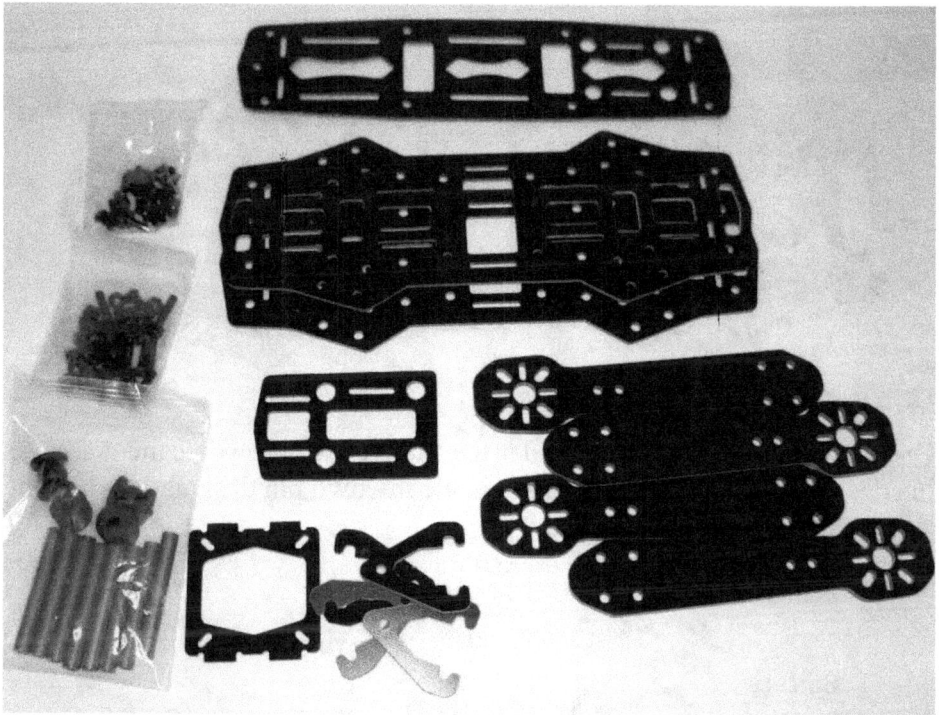

To assemble the 250mm quad, start by attaching the 8 red metal spacers to one of the bottom pieces. You will also need to attach spacers for the power distribution board. Then the two bottom pieces bolt together with the screws going through the legs in the middle.

The frame kit did not come with a power distribution board so you can make your own out of a 1.5 inch by 1.5 inch circuit board that has copper on one side. Power distribution mounting holes are drilled on 1 3/16 inch centers. Then a 1/10 inch wide strip of copper is removed down the center of the board. If you are using a 4in1 ESC then the power distribution board is not needed.

Some versions of this quad come with a 4in1 ESC that mounts in the center and eliminates the need for a power distribution board. They also have a flight controller that mounts above the 4 in 1 ESC saving even more space. I decided to do the same thing by modifying the quad slightly.

To better fit the 4in1 ESC and the multiwii flight controller I drilled eight extra holes near the center of the base of the frame as you can see in the next picture. The inner holes are for the 4in1 ESC and the new outer holes are for the multiwii flight controller.

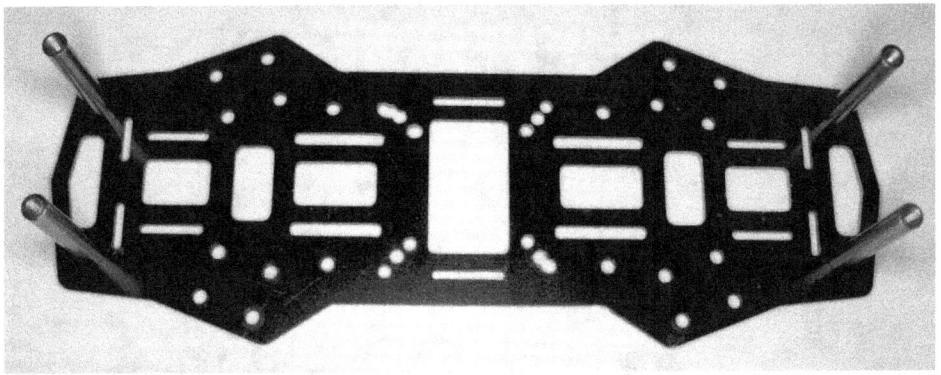

Once the 4in1 ESC is hidden under the flight controller you cannot see the pin designations. To resolve this, I made my own pin designation drawing.

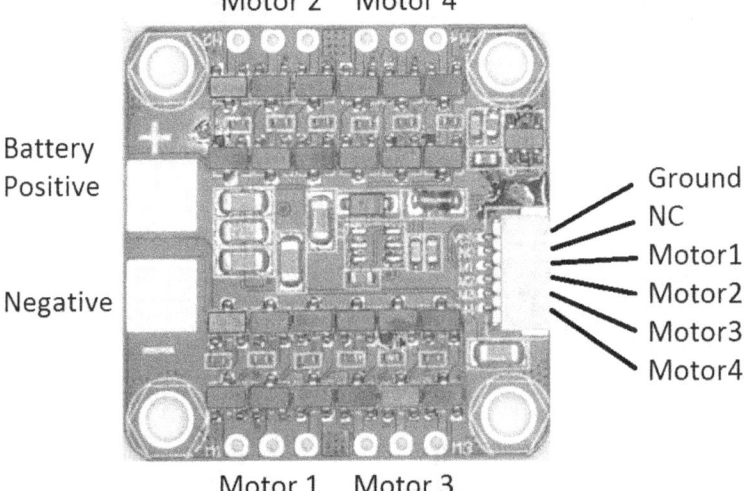

If you do not calibrate the 4in1 ESC it will pulse the motors when it is enabled. You cannot easily use the normal calibration procedure of plugging them in one at t time to the throttle of the remote receiver. The problem is there is no power coming from the ESC. You have to provide 5 volts from an AC adapter or some or a BEC.

This picture shows what the 4in1 ESC looks like with its wiring harness all soldered onto it. That harness takes 12 three and a half inch wires with bullet connectors and the power connections.

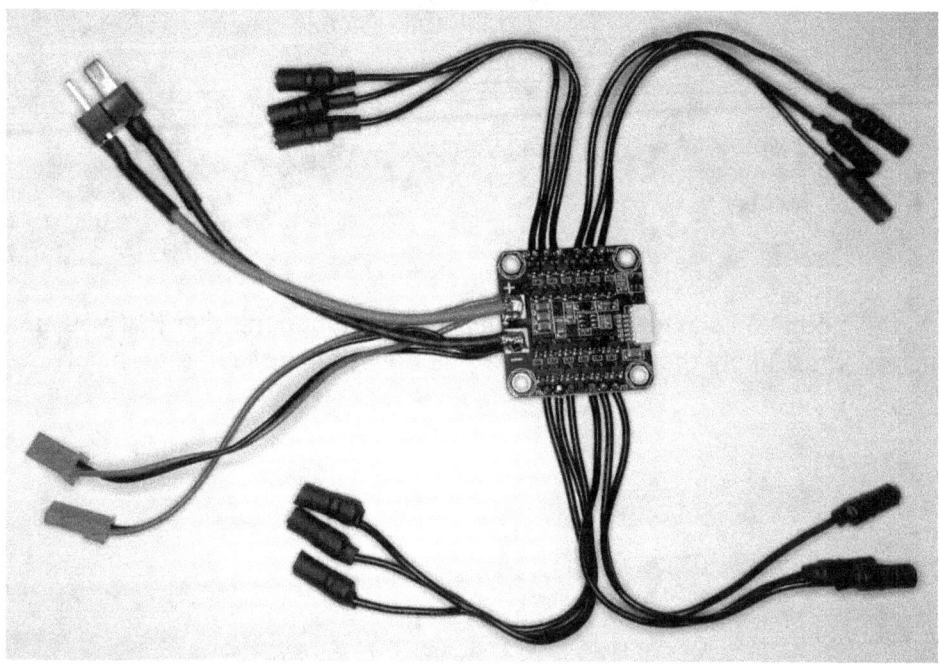

The next picture is what the 250mm quadcopter looks like once it is assembled with the top deck removed. The video transmitter goes on the top deck. If you have a smaller transmitter and a bigger battery the two can swap places. A 1500 mAh Battery is the biggest battery that fits in the lower deck area. A 16-gauge solid copper wire holds it in place.

The 4in1 ESC is located under the Multiwii flight controller. The 5V BEC is turned sideways and fits between the flight controller and the metal spacers. You will need two 12-volt power connectors for the 5V BEC and the video camera and video transmitter. You will need two 5V power connectors for the flight controller and for the LED's that go under the motors.

At this point you can go to the chapter on the flight controller that you are using. Most likely that is the multiwii as it is small enough to fit nicely.

Chapter 10

The F-550 HexaCopter

The F-550 HexaCopter is similar to the F-450 but it has six ESC's and six motors. This picture is of the F-550 frame advertisement on eBay. The F-550 frame kit runs about $25.

F550 550 PCB Board Frame Kit Hexacopter Drone For DJI Firewheel F550 Quadcopter
(112295437204)

Estimated delivery: **Sat, Jun 10 - Fri, Jun 30**
Tracking number: LK258682453CN

This picture shows what the F-550 frame kit looks like. There are six legs, two white, two black and two red. They are the same size as the F-450 legs. The central frame has a top and bottom. There are two bags of screws one is for the frame and the other is for mounting the motors.

Like on the F-450 you start assembling the frame by soldering the ESC wires to the bottom. Note that the power connections are reversed from all of the ESC's in polarity. I assume that the two sets of extra small slots designate the front and back of the Hexacopter. The smaller slots would be where you would mount a camera.

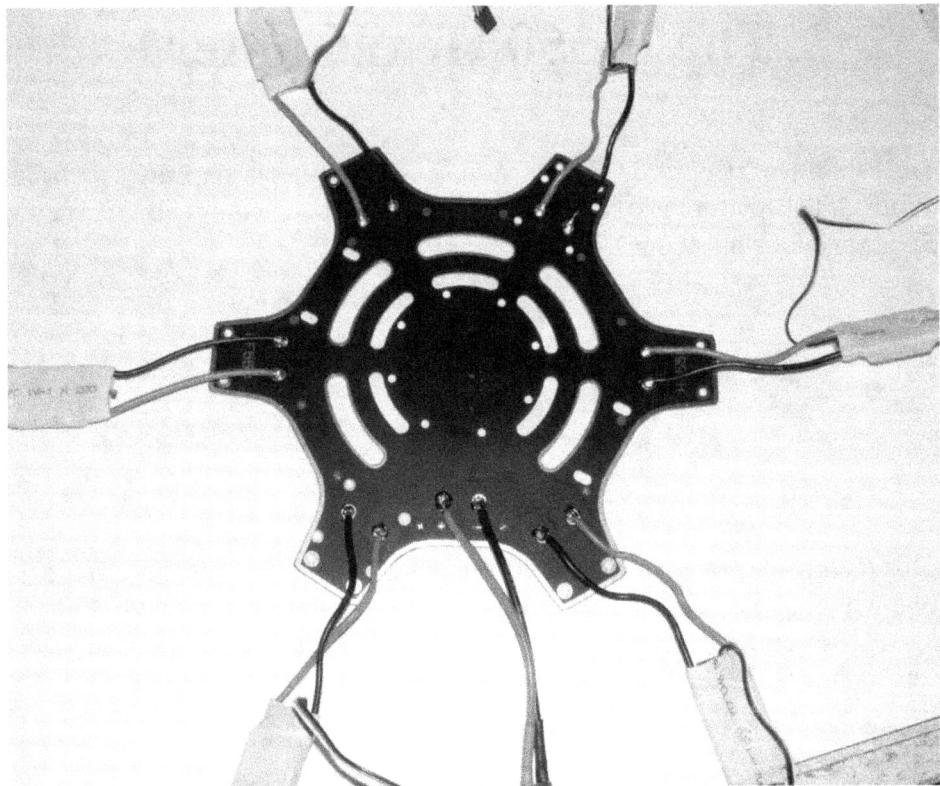

The top piece of the F-550 frame does not have holes in it for mounting the flight controller. From examining the holes, the flight controller was likely supposed to be mounted on the bottom frame. I prefer to mount the flight controller on top so it is easier to access to make changes.

Take a two inch square circuit board with holes on .1 inch centers and drill out the correct mounting holes at 1.4 inch centers for the MultiWii or other flight controller. This circuit board can then fit under the top piece. Plastic spacers that are used to hold the MultiWii will lock the board in place and will fit up through the slots in the top frame piece. This adapter can also work with the Nano/Uno adapter that was discussed earlier. With the two adapters you can mount a Nano or Uno on a Hexacopter as you can see in the next picture.

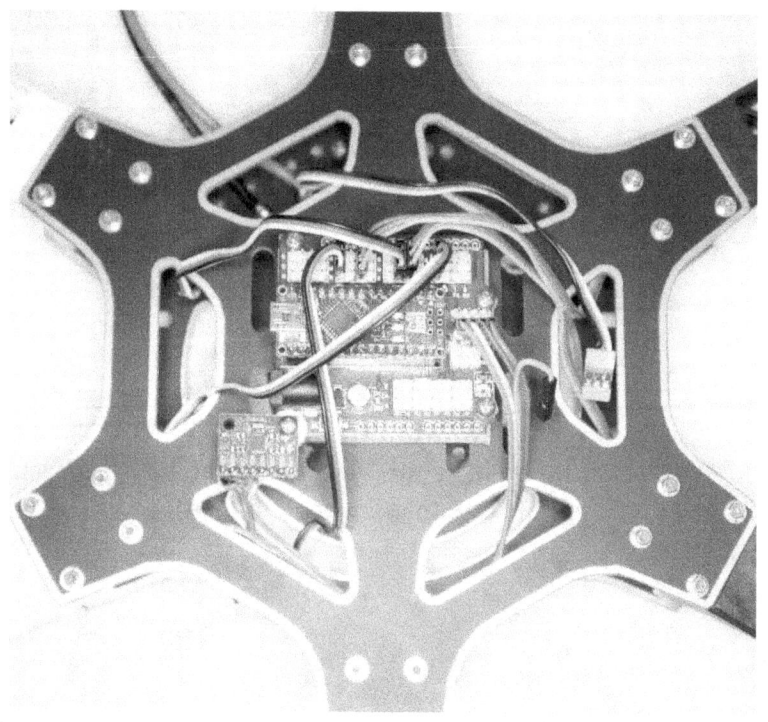

Getting the HexaCopter working is not as easy as removing the slashes in front of "#define HEX6X" in the MultiWii setup. The default HexaCopter settings will steal D5 and D6 from the remote control receiver and dedicate them to running the two extra motors instead. You are supposed to use a PPM remote receiver that can send all four remote controls over one cable.

To fix that issue you have to find a line in config.h where it can switch to using A0 and A1 instead of D5 and D6. For some reason this is found under "ProMini Specific Settings" about 1/3 the way through the config.h file.

However even then the hexacopter was not being properly controlled. Although the software now shows that the remote controls are reaching the flight controller and all six propellers work with the throttle, they still do not respond to the right, left, forward and backward commands. Flying without the proper controls is not possible. You might have to hold the hexacopter in your hand with the throttle on the low side to see if it responds properly to the remote control.

The next picture shows the code to replace D5 and D5 with A0 and A1.

```
................ ............
.......        Firmware Specific Settings      ............
................................................

.............................  Hexa Motor 5 & 6 Pins  ............
/* PIN A0 and A1 instead of PIN D5 & D6 for 6 motors config and promini config
   This mod allow the use of a standard receiver on a pro mini
   (no need to use a PPM sum receiver) */
#define A0_A1_PIN_HEX

.............................  Aux 2 Pin  ............
/* possibility to use PIN8 or PIN12 as the AUX2 RC input (only one, not both)
   it deactivates in this case the POWER PIN (pin 12) or the BUZZER PIN (pin 8) */
//#define RCAUXPIN8
//#define RCAUXPIN12
```

The multiwii uses the following configuration for the ESC's. Basically it adds the two middle motors as either D5 and D6 or A0 and A1.

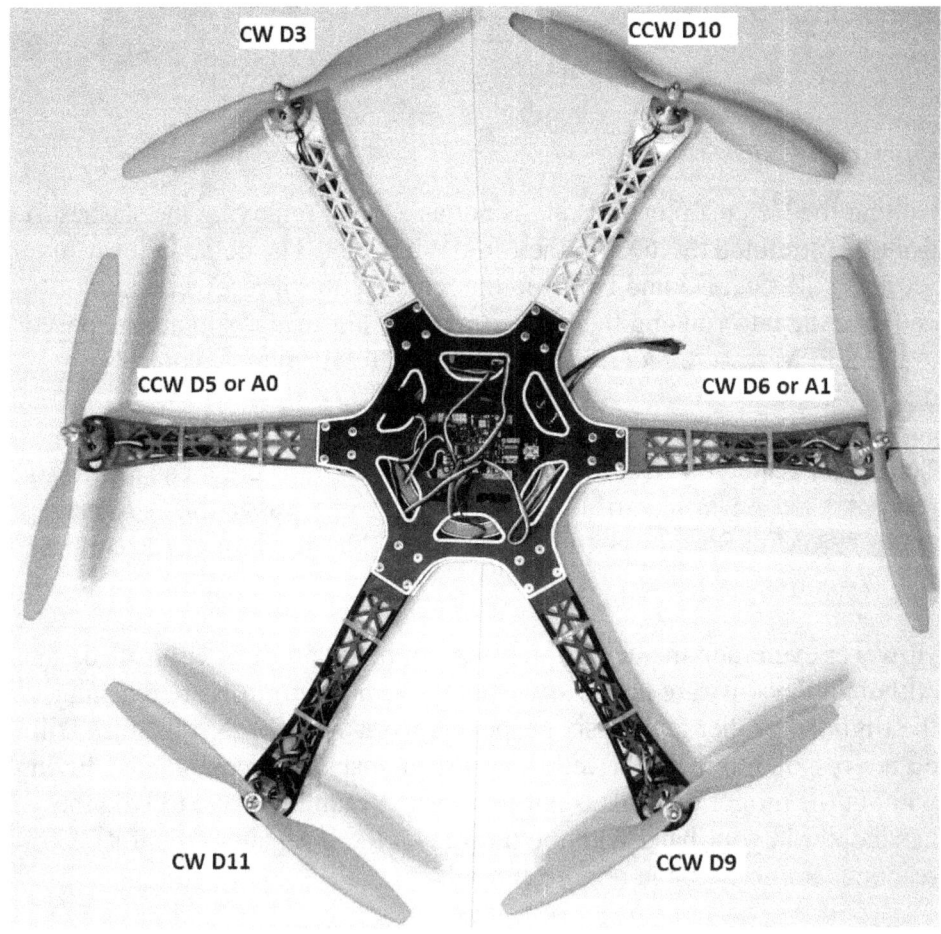

For the HexaCopter I made my own propeller protector. It is three feet in diameter. I intentionally designed it to look a lot like a UFO. It is made out of thin plywood or other thin light wood. You can add lights around the outside edge for an even greater UFO effect. Here is the design and dimensions for the propeller protector.

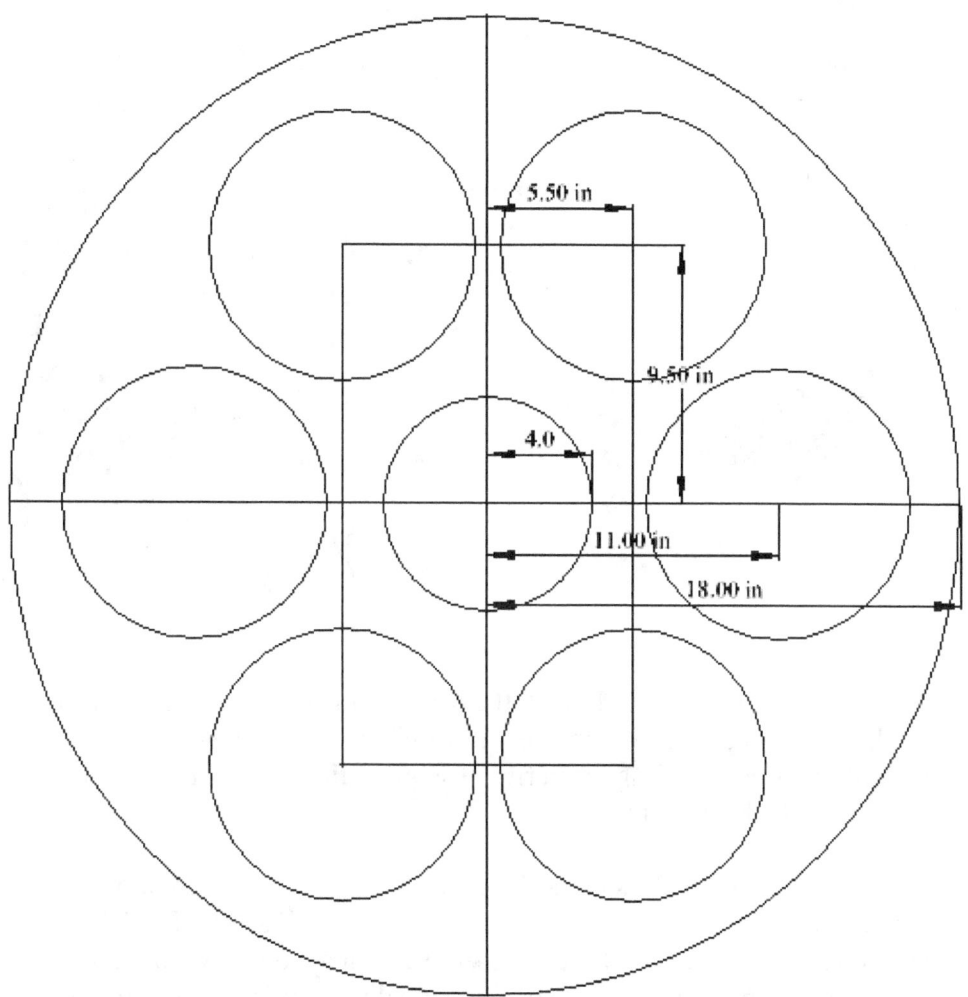

I had some four feet by two feet wide sheets of leftover luan. So I made my propeller protector as two identical sides that fasten together. After cutting it out with a jig saw I spray painted it gloss black. It took three coats of spray paint. This next picture is what it looks like assembled with the blade protector installed. I made the center hole a little bigger after this picture was taken.

After having a lot of trouble with the MultiWii I tried the APM controller on the hexacopter. However, the APM refused to arm every time claiming the GPS was not working properly. It worked fine every time on the quad but not on the hexacopter.

Finally, I tried the HobyKing KK2 controller and it works great with the hexacopter. The setup is easy and the controls work great. The KK2 controller uses motors that are numbered 1-6 in a clockwise rotation. However, the motors rotate in the opposite direction from the MultiWii setup.

Chapter 11

Brushed 105mm Quad

Just in case you might be inclined to build one of those tiny brushed quadcopters this chapter will cover that option. Usually brushed quads run on just one battery at 3.7 volts. That voltage eliminates the need for BEC's or voltage regulators. There are not a lot of multiwii brushed flight controllers available on eBay. In fact, I could only find one flight controller that was multiwii based.

First you will need a frame, here is the frame that I choose. It goes for about $9 on eBay. I did not use the upper deck as it was not necessary. Some people put the upper deck below the frame to mount stuff down there away from the propellers.

LANTIAN 105 105mm Carbon Fiber DIY Micro FPV RC Quadcopter Frame Support 8520 Co
(182305500257)

Estimated delivery Mon, Mar 20 - Wed, May 03

Next you will need some brushed motors. However, these 7mm motors were too small for the frame. They come in 7mm and 8 mm sizes. Three or four layers of electrical tape can fix the size problem. The price for the four motors was only $5. These came with connectors that match the flight controllers jacks.

SunFounder 4pcs 3.7V 7x20mm DC Coreless Motor for 6D-BOX Quadcopter
(371762530736)

Estimated delivery Wed, Nov 08 - Fri, Nov 24
Tracking number: LS092398538CN

You will need some small propellers with a one mm center hole to tightly fit on the motor shaft. Depending on your arrangement you can use two inch or three inch propellers. These were 20 for $5 but are a bit on the small size at 1.75 inches. However they will clear most electronics and the upper frame part.

20pcs 3-Blade Propellers for 7mm 8.5x20mm Coreless Motor RC Quadcopter
(381845228073)

Delivered on **Nov 15, 2017**
Tracking number **74899992140434458171**

Next you will need some batteries. They do not last for very long so having two of them is a minimum. This deal was for four of them at $7 with the charger.

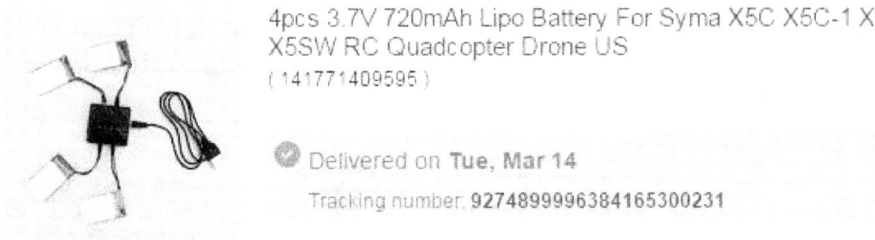

4pcs 3.7V 720mAh Lipo Battery For Syma X5C X5C-1 X5 X5SW RC Quadcopter Drone US
(141771409595)

Delivered on **Tue, Mar 14**
Tracking number: **92748999996384165300231**

Of course you will also need a flight controller. This is the only multiwii compatible flight controller that I could find on eBay. It runs about $12. The flight controller is called the "Sunfounder 6D-Box" flight controller.

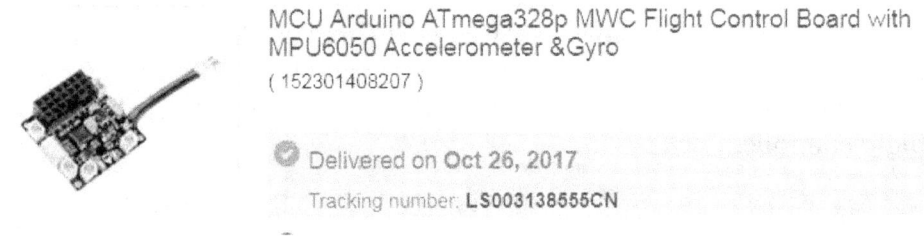

MCU Arduino ATmega328p MWC Flight Control Board with MPU6050 Accelerometer &Gyro
(152301408207)

Delivered on **Oct 26, 2017**
Tracking number: **LS003138555CN**

This flight controller is really really tiny. The holes are even too small for most mounting screws. You can use double sided sticky foam to mount it in place. Here is the flight controllers size compared to a quarter.

The remote control receiver mounts right on top of the flight controller. However, there is a jumper to modify the channel assignments. The jumper only changes one channel assignment. There will be more on this later.

You will also need an adapter to connect the flight controller to your PC for programming. The one in the next picture matched the order of the pins exactly. The flight controller has a six-pin connector but only five pins are actually used.

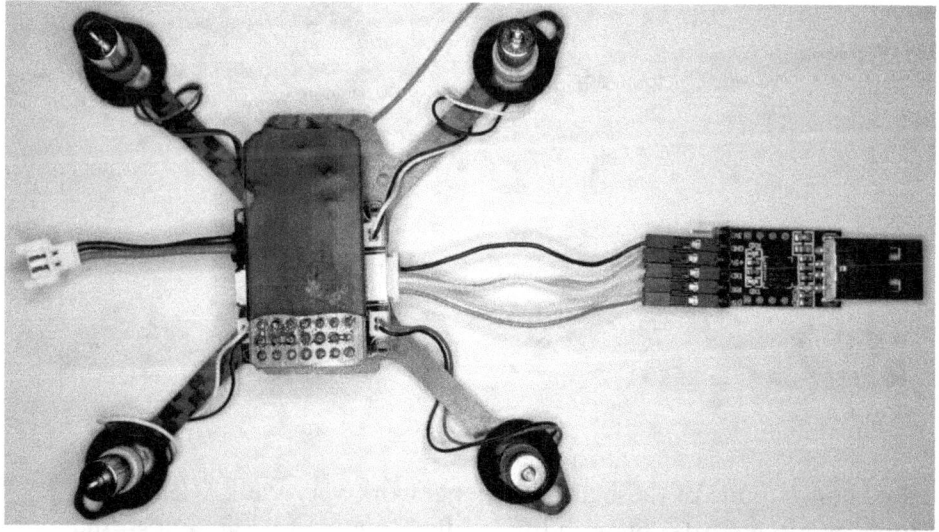

The serial Pinout from the front:
1. Not used
2. Ground

3. Power
4. TXD
5. RXD
6. DTR

In the multiwii config.h the flight controller to select is the "Sunfounder SE".

There are two problems with mounting the remote receiver right on top of the flight controller. One problem is the propellers are limited to two inches in size. Problem two is that the channels are reversed from what they should have been. That can be corrected in software according to the sunfounder manual. However it can also be corrected by mounting the remote control receiver under the frame. You will need six or eight one inch male to female jumpers to connect it to the flight controller. This arrangement allows for the use of three inch propellers as seen in the next picture.

To be honest none of these arrangements work very well. The tine propellers do not offer enough lift and the 7mm motors are not strong enough for the three inch propellers.

Chapter 12

Some Video Options

There are two methods of shooting a video of your flight. One method is using an on board camera that records everything. The second method is to broadcast the video back to the operator. The simplest and cheapest on-board camera is a car "Dash Cam". They are available for less than $15 on eBay. I picked up some at Wal-Mart during a sale for $9 each.

A Typical dash cam requires five volts regulated on the square-ish USB connector. You might have an older USB charger around with the right connector on it. The USB connector will need to be adapted to a two-pin connector. The ESC's provide five volts to the Flight controller. You can tap into that power by using the power and ground pins of an unused three pin connector.

The dash cam needs to be programmed for five or 10 minutes of recording time and then strapped onto the front of the quadcopter. Use heavy duty straps or rubber bands to hold it in place. If connected right it will power up when that battery is connected to the drone and power down shortly after power is removed.

Here is the above dash camera ad as it appears on eBay. It comes with a suction cup mount and cigarette lighter adapter. They go for about $12. They are really easy to set up as they automatically come up in video record mode and shut down shortly after power is removed.

1080P Car DVR Camera Dash Cam Video 2.3" LCD G-sensor Night Vision Necessary HR
(252650335868)

Estimated delivery **Tue, Sep 26 - Mon, Oct 16**

This item has been shipped.

Besides the dash camera shown in the previous picture there are other dash camera options. One is to use a dash camera with a fold up screen. This model runs about $12. It has a battery compartment so it can run independently of the drones power. However, this camera is rather large for use in smaller quads.

2.5" Full HD1080P 6LED Car DVR Vehicle Camera Recorder Dash Cam 270° IRSS MC
(162144532345)

Option: Recorder

Estimated delivery **Wed, Aug 16 - Thu, Aug 31**

There are what are called "Sports Cameras" that come with lots of mounting options. They are knock off of a popular camera. They even include a waterproof cover. This one was only $14 to buy. They can be mounted on a drone with two rubber bands or some double sided sticky foam tape.

2.0" HD 1080P Helmet Sports Action Video 30M Waterproof Camera DV No Wifi SJ4000
(253154710419)

Delivered on **Nov 03, 2017**
Tracking number: **LK372556839CN**

Another video option is to skip the screen completely. This makes for a rally small camera for only $6! This camera is however hard to configure with only one button and no screen.

Wide Angle FPV CCD 600TVL/700TVL Mini CMOS Camera Camcorder DVR RC Aircraft Spy
(172592889165)
Model: Mini DVR Camera

Delivered on Thu, Aug 03
Tracking number: LK302513553CN

Some of these cameras are so small that mounting them can be an issue. There is a solution for problem that as well. It only costs about $1 from China. The camera angle is easy to adjust.

FPV Racing Drone Camera Tilt Mount Universal Lens Adjustable Black Supporter LN
(381756263031)

Estimated delivery Wed, Aug 16 - Wed, Sep 20
Tracking number: 69109279015

The next video option is to transmit the video back to you so you can see what the camera sees. This is called FPV or "First Person Video". With this addition you could fly the quadcopter through tight spaces and see what is happening live on a screen. You can get a complete wireless camera transmitter and receiver kit on eBay for around $25. What you will need is the camera, power adapter, cables, a receiver and a composite monitor to see the results on.

Up next is a picture of a typical wireless security camera, with six IR night vision LED's. It can take a fresh new nine volt battery to power the camera for about an hour. You can tap into the quadcopter battery for a power source as well.

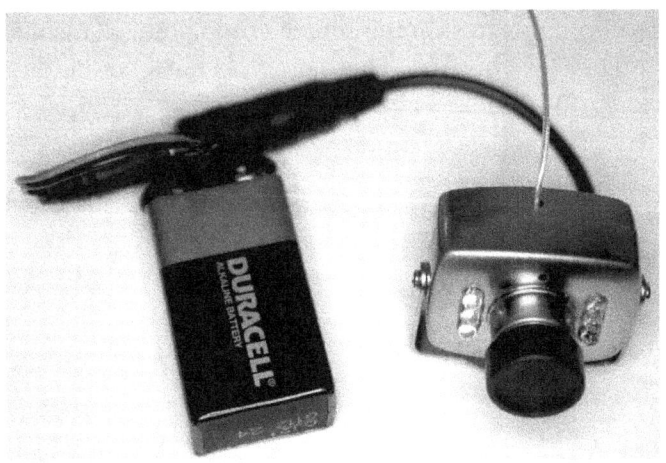

Up next is a picture of the quadcopter with the wireless camera mounted on the front. You could even add some servo motors so you would be able to rotate the camera right and left or up and down.

This next picture is of the wireless receiver. It uses a 9 to 12 volt DC adapter for power. There is a silver frequency knob that has to be tuned for the best picture quality. The audio and video out use RCA cables to go to your screen. You can also connect the RCA cables to a USB video adapter and then you can watch what is happening on a computer screen.

The resulting picture quality is not that great. The cheap camera lacks good low level light sensitivity. Up next is a typical screen shot from a TV attached to the receiver using normal room lighting. You can see the lack of picture quality when just using typical room lighting. It should do much better in bright sunlight. A general rule is that the more expensive the camera the better it does in low lighting conditions.

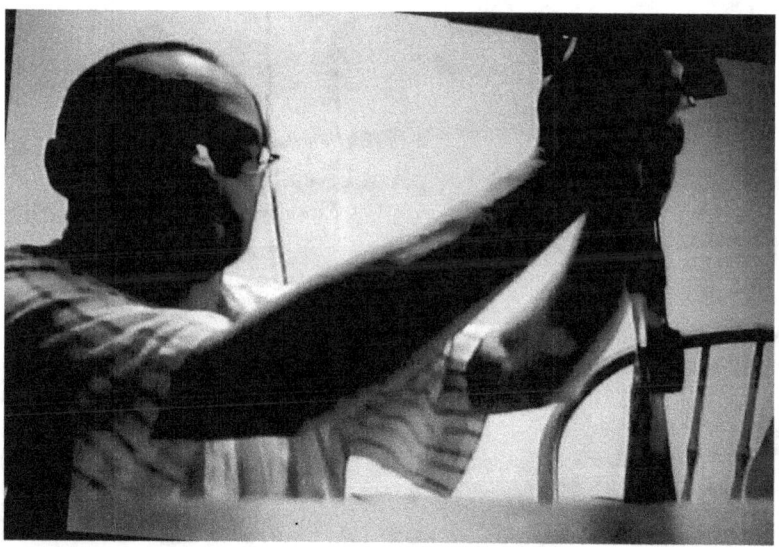

Cameras with built in transmitters are usually limited to 25mw. If you buy a camera with a separate transmitter you can have more power and you can also add an "On Screen Display" or OSD module. That will allow you to remotely see the status of several parameters like the battery. This picture shows a camera with a 200mw transmitter and a receiver.

A cheap camera is only about $5 without a case. Things like a wide angle lens and better sensitivity usually costs a little more. So more expensive cameras are better in dim lighting. The light sensitivity is measured in "Lux". I have seen cameras on eBay that vary from a minimum of 1 lux sensitivity to a very sensitive .0001 lux or more.

 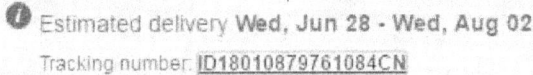

The transmitter and receiver normally run about $25 together but by bidding on a set I was able to get them both for only about $20.

Wireless Video Receiver & Transmit System TS832 & RC832
(162544254113)

Delivered on **Thu, Jun 15**
Tracking number: 9400109699939021471643
This item has been shipped.

Of course while we are at it, lets add the OSD module. It runs about $8. It only works with a flight controller with a "telemetry" output. The labels on the side of the plastic box furthest from the pins do not match up with the pins. It is best to get your camera and transmitter working first then add the OSD.

Mini OSD Rev. 1.1 OSD On Screen Display DIY Drones Ardupilot APM2.6 APM2.5
(121538997640)

Estimated delivery **Thu, Jun 29 - Thu, Jul 20**
Tracking number: LK276126695CN

The mini OSD does not have a battery monitor built in however there is a hack on line to add a battery monitor if your flight controller does not have one. Here is a schematic to show how to hook up the last three items.

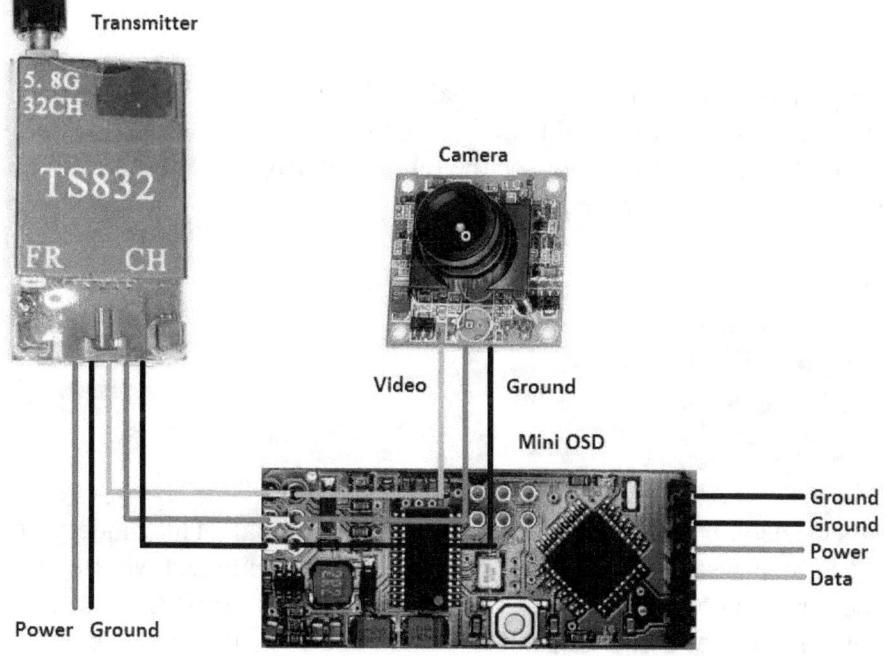

If space is an issue there is a much smaller OSD. You have to connect your own wires to it and it does not have a voltage regulator so everything has to be five volts. It only costs about $5 on eBay. However it has two battery monitors if your flight controller does not have one.

But in order to see the status of the battery the flight controller has to have a tap into the battery circuit. This one fits the APM and has a .01 ohm resistor so it can monitor the current as well as the voltage. It runs about $5 on eBay. Make sure the plug matches your batteries plug.

While you are at it, you can add Virtual Reality Goggles! With these you can see the screen in the bright sunlight but you will have to operate the controls "blind" as you cannot see the remote controller. These run from about $50 to several hundred dollars.

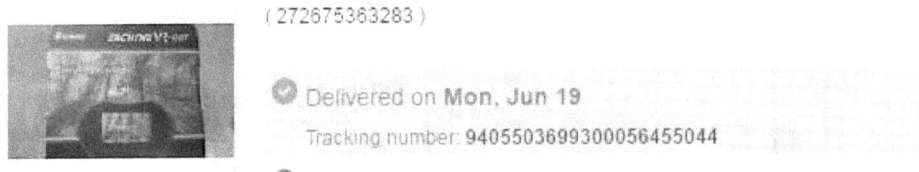

Another video option is to add a pan/tilt camera gimbal. This model is a little tricky to assemble. The trick is to use a sharp knife and whittle the plastic arms that come with the servos so that they will fit in the spaces that are allotted for them. Also, it is designed to fit some unknown model of camera. However, by using some double sided tape any board type camera will fit. This gimbal sells for only about $2 without the servos. I have about a dozen small servos from RC airplanes so finding the servos was not a problem.

PT Pan/Tilt Camera Platform Anti-Vibration Camera Servo Mount for Aircraft FPV I
(390961293160)

Estimated delivery **Tue, Aug 08 - Tue, Sep 12**

This item has been shipped

This next picture shows what it looks like once it has been assembled.

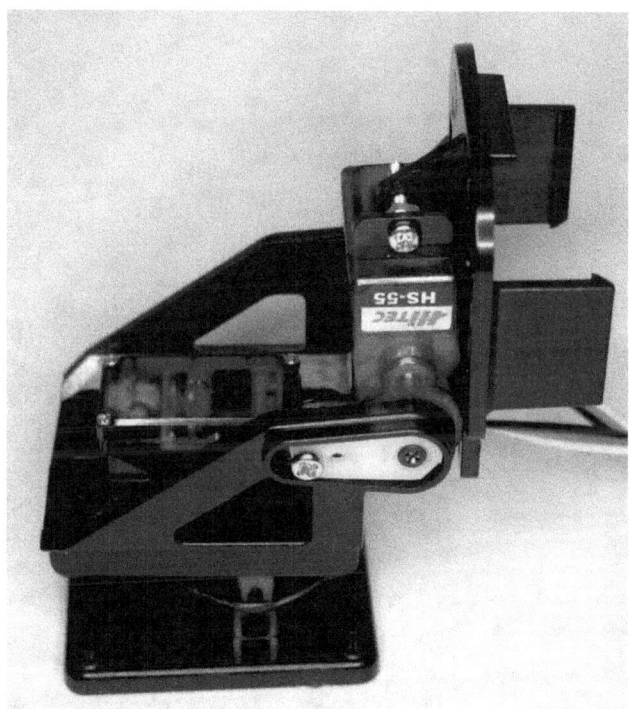

To run the two servos in the gimbal you will either have to obtain two channels from the main remote receiver or use a second remote transmitter and receiver.

Chapter 13

Optional Upgrades

You might want to pimp out or add some fancy features to your quadcopter. In this chapter I will introduce you to some of these available options and let you know what I think about them.

Landing Gears help reduce the shock of landing on the legs. They come with mounting screws because the screws that came with the Quadcopter are a little to short. These are cheap at 4 for $2.

DJI Wheels Tripod Landing Skids Gear for F450 F550 SK480 Aircraft Qudcopter DP
(112208321244)

Estimated delivery **Mon, May 22 - Wed, Jun 14**

This item has been shipped.

The landing gears make the drone a lot taller. I tried flying in a slight cross wind and the drone had a tendency to tip over during take off and landing because the landing gears are too close to the center.

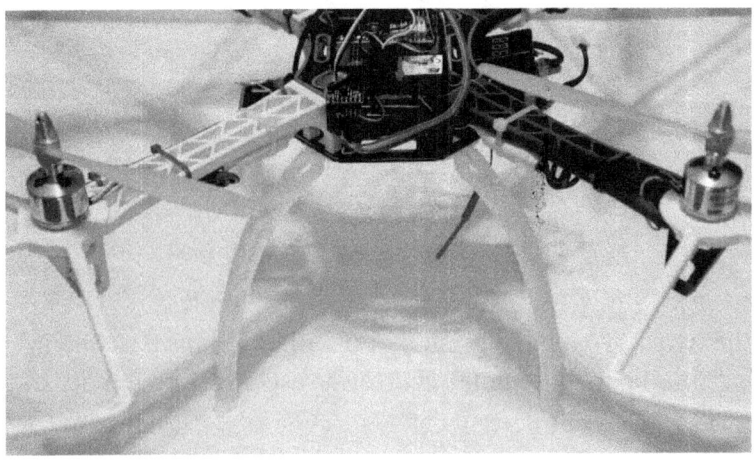

Landing skids are a better option. They are longer and hence reduce the tipping over problem. This assembly costs about $10.

DJI F450 F550 Frame Landing Gear Landing Skid FPV Aerial Photography
(182556084445)

Estimated delivery **Tue, Jul 18 - Mon, Aug 07**
Tracking number: LK292968133CN

A Battery Monitor plugs into battery charging connector and lets you know when to recharge the battery. It also gives you the status of all of the individual cells in the battery. It runs about $2.

1-8S BX100 Lipo Battery Voltage Tester/ Low Voltage Buzzer Alarm with Dual Speak
(172596181477)

Estimated delivery **Thu, May 18 - Fri, Jun 23**
Tracking number: ID18010682964141CN

The battery monitor plugs in with the black wire(s) towards the left side. It needs to be mounted somewhere near to the battery for that reason.

If you are not an expert flyer you will need some spare propellers. I broke 4 propeller blades in the first week of flight. Five pairs runs about $8.

5 Pairs Nylon 1045 1045R CW CCW Propeller Orange for DJI F450 500 F550 Multi US
(131512487670)

Delivered on **Fri, Apr 28**
Tracking number: 9374869903501916296021

Propeller Guards like these should reduce the number of broken blades. They come with mounting screws to go through them, the arms and into the motors. Four of them run about $8 on eBay.

4Pcs High Quality Propeller Protector For DJI F450 F550 RC Quadcopter US H5X8
(332155928764)

Estimated delivery **Thu, May 04 - Fri, May 12**
Tracking number: 9374869903501932244433

The addition of these propeller protectors increases the width from 18 inches (450mm) to 31 inches. They are like training wheels for your Quad-Copter. They are breakable as I have found out.

A better option is removable propeller guards. They can be snapped off for ease of transporting the Quad. However, they are more expensive at $22 for six of them.

Snap On/off Tool-Free 6x White Prop Guards for DJI Flamewheel F550 H3-3D HERO4
(281558032796)

ⓘ Estimated delivery **Thu, Jul 20**

You can also add "running Lights". They make LED strips that are up to 16 feet long. However, they can be cut down to 6 inch strips. They also come in 5 volt and 12 volt versions. Mine are 12 volts so they have to be connected to the battery. I added a two-pin connector to where the battery connects to the main bottom board and then a matching 2 pin connector is soldered to the LED strips.

The LED strips need to be underneath the wire ties. They come with sticky strips on them but that will not hold up without the wire ties. It is best to solder wires to the LED strips before attaching them to the Quad. One of the LED strips can be seen in the next picture.

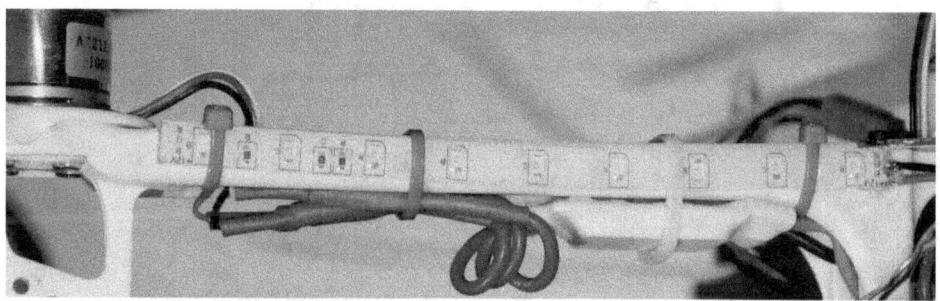

Another option for lights are small circuit boards that sit under the motors. They are about $1 each from China. However, the model shown below does not fit any motors! Also, these do not fit the F-450 or F-550 motor mounts. Look for a model with slots for the motor screws instead!

Lightweight Matek RGB LED Circle Board 7-colors X8 16V For FPV RC Multicopter
(252911611605)

Quantity: 2

ⓘ Estimated delivery **Wed, Jun 14 - Wed, Jul 19**
Tracking number: SYBAB38844012

This model has slots to fit the motor mounts but it has less LED's (only 2) as well. They are about $1 each. You should really have two red and two green ones to designate the right and left side of the quad.

2pcs LED CIRCLE X2/5V Round 1806 2204 2206 Motor Mount Light Board for FPV Drone
(252603480716)

Color: 2pcs Green

Estimated delivery **Thu, Jul 27 - Wed, Aug 30**

Tracking number: ID180110118974960CN

Here is what the LED's look like mounted on the bottom of the ZMR250 Quadcopter.

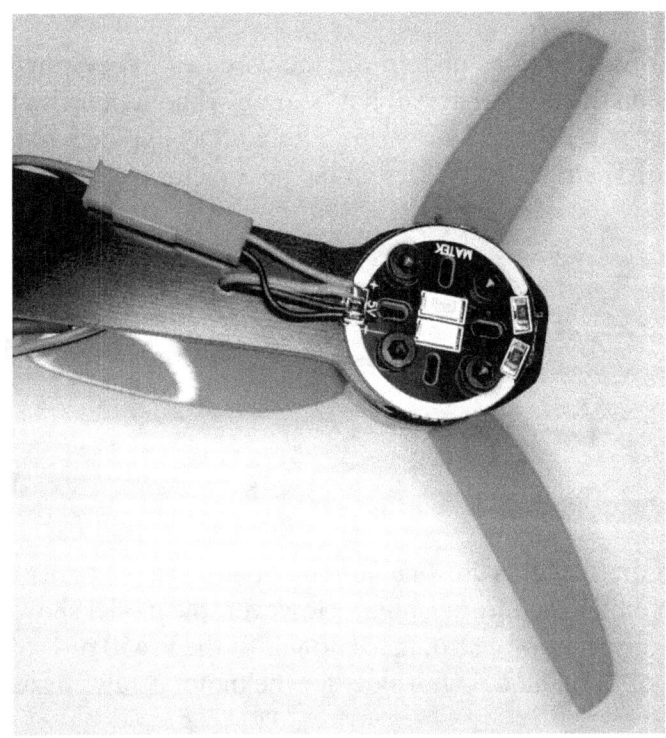

Want to make a bigger quadcopter or hexacopter with even longer propellers or some bigger motors? You can buy or make motor extension plates. These extenders bolt unto the existing motor mounts and have holes in them for bigger motors.

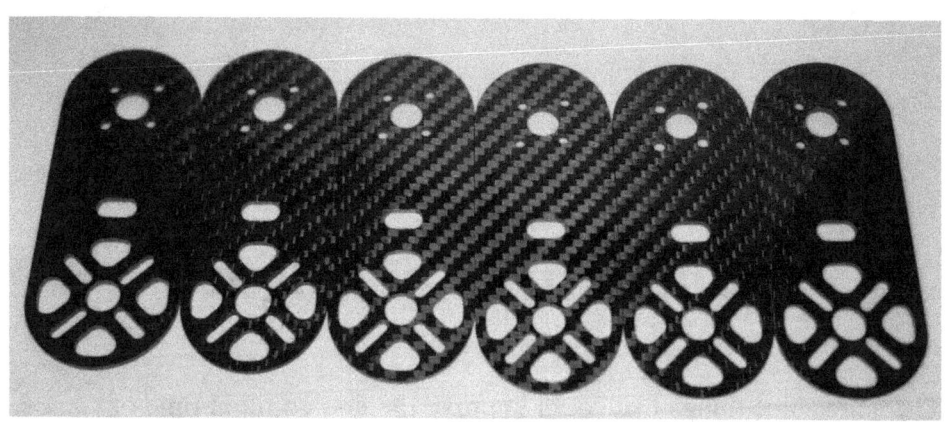

Chapter 14

The Future of Flight

There are some things that could improve the limited flight time of quadcopters. One big issue is that when they are flying forward they are typically on a 45 degree angle. This creates lots of wind resistance on top and drag underneath them. One solution is to tilt the rotors so they angle up to 90 degrees forward and in some cases backwards. This is called a "tilt rotor" quad. There are two frame kits that have these tilt rotors. They are the VQ250 and the SL300 and are available on eBay for about $55-$65. Here is a typical eBay ad.

Tilt Rotor FPV Quadcopter VQ250 Frame Kit
Brand New
$49.00 From Canada
Buy It Now
+$9.00 shipping
13 Watching

There is actually a full size flying plane with tilt rotors. It is called the Osprey V-22. One of the biggest problems it had to overcome is the transition from vertical to horizontal flight. There is also an assembled three propeller tilt rotor remote control aircraft with VTOL. It is the E-Flite EFL11050.

E-flite Convergence VTOL BNF Basic, 650mm, EFL11050
Brand New
★ ★ ★ ★ ½   13 product ratings
$219.99
Trending at $234.97
Buy It Now
Free Shipping
8 watching
13 New & refurbished from $219.99
1 Pre-Owned from $215.00

Convergence VTOL (Vertical Takeoff Landing) Bind N Fly Basic E-Flite EFL11050
Brand New
★ ★ ★ ★ ½   13 product ratings
$219.99
Trending at $234.97
Buy It Now
Free Shipping
20 sold
13 New & refurbished from $219.99
1 Pre-Owned from $215.00

Basically by tilting the rotors forward and adding some sort of wings you can as much as triple your flight time.

Another possible improvement to flight is ionic flight. Supposedly the B2 bomber can run in ionic flight mode and be totally quiet as it flies. Basically you add a wire in front of the wing and charge it up to several thousand volts. This cuts through the air better and reduces the turbulence caused by jet engines or propellers.

There is an experimental ionic lifter that consists of a wire suspended above aluminum foil wings. Usually these are triangular shaped and only lift a short distance as they are tethered to a power supply. However, by itself it cannot lift even its own power supply. Ionic lift works best in conjunction with other types of propulsion.

So how would you put a wire in front of the "wings" of a quadcopter? For one thing the propellers are the wings. Fastening a wire to the front edges of the propellers is not n easy thing to do. An alternative is to fasten a wire above the propeller and an aluminum collector below the propeller. Then apply 20 KV to the wire and see if the turbulence is reduced.

Issues you might experience include arcing to the motor shaft, and to the ESC's. In fact, I have toasted one ESC. Make sure the motor frame is grounded so that any arcs go to the aircraft ground not into the electronics.

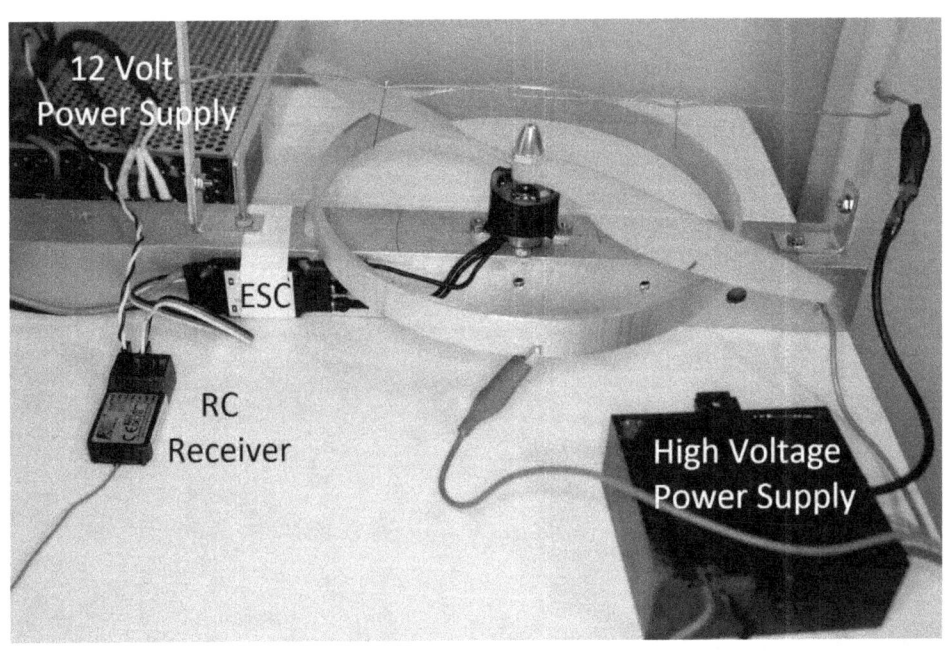

Chapter 15

Definition of Terms

Accelerometer – Device to measure the amount of acceleration.

AGL – Above Ground Level – Your altitude or distance above the ground.

APM – AutoPilot Mega – Flight Controller based on the Arduino Mega. Features a larger and more powerful controller than the MultiWii.

Arduino – Open Source small processor and circuit board used to control anything from LED's to Drones. Includes the Uno, Nano, Mini and many other variations.

ARF –Almost ready to Fly – Usually denotes a kit that contains most of the parts needed to build a drone.

Autopilot or Autonomous Flight – When an unmanned flying vehicle can fly without anyone at the controls. Usually uses GPS to find its destination.

Balanced Battery Charger – A battery charger for multi cell Lithium Polymer batteries. Needed to properly charge the individual cells.

Barometric Pressure Sensor - Used to determine altitude. Necessary for altitude hold functions.

BEC – Battery Eliminator Circuit – uses voltage regulators to provide additional voltages without additional batteries.

Bind – Connecting a transmitter to a receiver. Being on the same frequency is not enough you have to match up the codes as well.

Brushed and Brushless Motors – In a brushed motor the coils rotate and contacts deliver the power to the coils. In a Brushless motor the coils are fixed and the magnets rotate around the coils.

CF - Carbon Fiber – A very durable yet light weight material used in building aircraft.

Drone – A generic term that can refer to any unmanned flying vehicle. Usually one that can fly on autopilot.

ESC - Electronic Speed Control - Converts Servo Pulses into three phase motor speed pulses

FPV - First Person View – Video is transmitted back to the operator so they see what is happening just like they were in the flying device.

FTDI – Future Tech Devices International – USB to serial data converter.

Gimbal - A camera mount that uses servos to pan and tilt. Allows camera to stay on one object as the drone moves.

GPS – Global Positioning by Satellite – Uses Satellites to determine the present position or to find a destination.

Gyroscope – Used to be rotating devices that can detect movement to stabilize a vehicle. Now it is done with IC's.

Hexacopter – Similar to a Quadcopter but having six motors and propellers. It can operate without one of its motors.

I2C – Communications buss that allows several devices to communicate over the same buss.

IMU – Inertia Measurement Unit – Usually refers to a combination of a gyroscope and an accelerometer.

KV – this usually means kilovolts or a thousand volts but when it comes to multicopter motors it stands for Thousands of RPM Per Volt.

LiPo - Lithium Polymer battery – Light weight battery that is used in most drones.

mAh – milli ampere hour – measurement of power (amperes) over time (hours) a 2200mAh will deliver 2.2 amps for an hour or 22 amps for 6 minutes.

Motors – Rated in KV for Thousand RPM per Volt. A 1000 KV motor will reach 10,000 RPM at 10 volts.

Multicopter – a helicopter like flying device that can have 3 to 8 or more motors and propellers. The multiple rotors eliminate the need for a tail rotor to stabilize the flight.

MultiWii – A flight controller based on the Arduino and the motion detectors from the Wii game controllers.

NAZA – A popular flight controller.

Octocopter – Similar to a Quadcopter but having eight motors and propellers. Like the Hexacopter it can fly without one or more motors.

OSD – On Screen Display – Adds text to the FPV system to give the status of various things to the viewer.

Pitch – The flight angle that tells if it is headed up or down.

Propellers – Rated in Inches and Degrees of tilt. A 1045 propeller is 10 inches long with a 45 degree tilt. A 5030 propeller is 5 inches long with a 30 degree tilt.

PWM or PCM - Pulse Width or Code Modulation – Servo controls use the width of the pulse to determine the value. 1.0 ms is the minimum and 2.0 ms is the maximum.

PPM – Pulse Position Modulation – Sends multiple channels of PWM over one signal wire.

Quadcopter – A flying device similar to a helicopter but having four rotating propellers.

R/C - Remote Control or Radio Control – Takes a remote transmitter with controls that move in four (or more) directions and a receiver that applies those controls to the device being controlled.

Roll – Tilt to the left or the right, and hence go in that direction.

RTF – Ready to Fly – Indicates that the kit contains everything needed to make a flying device. A RTF device may not include the batteries.

Servo – short for servo motor – a motor that responds to pulse width signals usually contains a position sensor.

Throttle – Control that sets the speed of the motors.

UAV – Unmanned Aerial Vehicle

Ultrasonic – Sensor using high frequency sound for shorter distances such as height above the ground in feet.

Yaw – Control for rotation of the center axis, in a quadcopter it is to rotate right or left.

Bibliography

MultiWii website:
http://www.multiwii.com/

APM website:
http://ardupilot.org/copter/

A good web site for APM info:
https://airhigh.wordpress.com/tag/flight-controller/

Hobby King for KK flight controllers.
https://hobbyking.com/en_us/multi-rotors-drones.html

USB to KK Programmer
http://www.fischl.de/usbasp/

KK Flashtool and lots of good informtion
http://lazyzero.de/en/modellbau/kkmulticopterflashtool

www.ingramcontent.com/pod-product-compliance
Lightning Source LLC
Chambersburg PA
CBHW082346220526
45470CB00008B/2659